Cockpit Monitoring and Alerting Systems

P.M. SATCHELL

*a*SHGATE

© P.M. Satchell, 1993

All rights reserved. No part of this publication may be reproduced, stored in a retrieval system, or transmitted in any form or by any means, electronic, mechanical, photocopying, recording, or otherwise without the prior permission of the publisher.

Published by
Ashgate
Ashgate Publishing Limited
Gower House
Croft Road
Aldershot
Hants GU11 3HR
England

Ashgate Publishing Company
Old Post Road
Brookfield
Vermont 05036
USA

A CIP catalogue record for this book is available from the British Library

ISBN 1 85742 109 4

Printed in Great Britain at the University Press, Cambridge

Contents

Figures and tables — viii

Preface — ix

1. Introduction — 1

Part A Monitoring Problems and Processes — 5

2. Automation, peripheralisation and error — 7
    2.1 Automation in aircraft — 7
    2.2 Peripheralisation — 10
    2.3 Peripheralisation effects — 10
    2.4 Human error — 15
    2.5 Peripheralisation error modulators — 18
    2.6 Responses to errors — 26
    2.7 Peripheralisation and monitoring — 29
    2.8 Summary — 30

3. CRM as a response to peripheralisation — 31
    3.1 CRM origins — 31
    3.2 CRM objectives and key concepts — 32
    3.3 CRM effectiveness — 35
    3.4 CRM, accident processes and peripheralisation — 38
    3.5 Automation, peripheralisation and CRM — 43
    3.6 Summary — 46

4. Stress and arousal in cockpits — 47
    4.1 Stress — 47
    4.2 Stress and arousal — 51
    4.3 Cockpit stressors — 53
    4.4 Stressor management — 55
    4.5 Summary — 57

5. Vigilance mechanisms — 59
   5.1 Vigilance theories — 59
   5.2 Role of arousal — 63
   5.3 Vigilance system components — 67
   5.4 Arousal system components — 73
   5.5 Arousal vigilance interactions — 78
   5.6 Summary — 79

6. Automation, peripheralisation, vigilance and stress — 81
   6.1 Peripheralisation-vigilance cycle — 81
   6.2 Stress and vigilance — 81
   6.3 Peripheralisation-stress cycle — 83
   6.4 Multiple positive feedback loops — 84
   6.5 Solutions — 85

Part B Monitoring, Measurement and Alerting Systems — 87

7. Vigilance measurement — 89
   7.1 Measurement issues — 89
   7.2 Electroencephalography — 91
   7.3 Heart rate — 94
   7.4 Skin conductance — 96
   7.5 Adrenalin — 97
   7.6 Others — 98
   7.7 Workload, effort, vigilance and arousal — 99
   7.8 Summary — 101

8. Human alerting systems
   8.1 Historical perspective — 103
   8.2 Road transport alerting systems — 104
   8.3 Rail transport alerting systems — 108
   8.4 Aircraft alerting systems — 109
   8.5 Alerting system classification — 110
   8.6 Cockpit systems under development — 115
   8.7 Summary — 116

9. The ideal alerting system — 119
   9.1 Established objections — 119
   9.2 Ideal system characteristics — 120

9.3	Vigilance measurement	121
9.4	Operator control	124
9.5	Measurement issues	126
9.6	Characteristics of a future system	130
9.7	Summary	133

Part C Monitoring Management 135

10. Monitoring management, interim and future changes 137
    10.1 Flight phase 137
    10.2 Breaking the loops 139
    10.3 Present cruise phase management 140
    10.4 Interim changes in the organisation 142
    10.5 Interim changes in the cockpit 143
    10.6 Future changes 148
    10.7 Alerting system in other flight phases 150
    10.8 Alerting system in other monitoring roles 151
    10.9 Summary 152

11. Conclusions 155

References 157

Index 176

Figures and tables

Figure 2.1 Primary cause factors in 738 aircraft hull loss accidents in the period, 1959-1989 — 16

Figure 2.2 Peripheralisation produces effects which are open to modulation by a variety of factors — 19

Figure 2.3 A scheme for the interaction between automation, peripheralisation, its effects and human error — 28

Table 3.1 Human factors accidents, 1972-1982 — 39

Table 3.2 Human factors accidents, 1983-1991 — 40

Figure 3.1 An optimistic view of the effectiveness of CRM programs on the interaction between automation, peripheralisation and human error — 44

Figure 3.2 A pessimistic view of the effectiveness of CRM programs on peripheralisation effects — 45

Figure 6.1 The interaction between human error, automation, peripheralisation and vigilance — 82

Figure 6.2 The peripheralisation-stress cycle — 84

Figure 9.1 Sweat output from the left index finger during two short duration simulator flights — 127

Figure 9.2 Sweat output from the left index finger while flying a simulated Beechcraft Starship south from Chicago O'Hare airport and return — 129

Preface

Some monitoring roles are more critical than others, the most important being those where monitoring failure can have disastrous consequences. Monitoring for such occurrences has become increasingly automated, particularly in the cockpit of commercial aircraft. The goals of cockpit monitoring are to maintain awareness of the state of the aircraft itself, the state of the aircraft as a transport machine in a potentially hostile environment, the state of passenger satisfaction and the state of the flight as a commercial enterprise. Automation is dramatically and continuously altering the character of these monitoring activities. The interaction between automated control systems and human monitoring prowess points to the need for a new approach to monitoring in the face of advancing automation, an approach where the monitoring abilities of humans are maximised rather than ignored.

I would like to thank Mr Ken Gould, Captain Dick Wilkinson and Dr Jack Woods for their encouragement. Captain Neil Alston, Captain Graham Beaumont and Mr John Hindley of Ashgate have provided invaluable advice on many issues related to this book. My wife Anne, and my daughters Amy and Olivia have been the ones that have made this book possible. Bob Martin has been the first to read the book and has made invaluable suggestions. I am grateful to all.

Paul Satchell
Sydney
August, 1992

1 Introduction

Monitoring is a very widespread human activity and requires the maintenance of appropriate vigilance levels. Vigilance is often critical to the individual, customers, the organisation and in some situations, mankind. In examining the monitoring role, only certain types of monitoring are considered, although there are implications for helping the alertness of all those in any monitoring role. The type of monitoring that is principally under consideration is that where the chance of an untoward event is low, where the untoward event may be new to the monitor's experience and where the consequences from an untoward event can be devastating. This includes the modern long range airliner cockpit, power stations (conventional and nuclear), aerospace applications, industrial monitoring of critical chemical and biological processes, many forms of transport, health care and financial market monitoring. Other critical monitoring areas occur in military and security systems. Monitoring might be seen as an outmoded activity, because it is a form of inspection of control systems which, if properly constructed, would be failure tolerant if not failure free. The capability of modern control systems does not allow such an optimistic view.

In this examination of alerting systems and the monitoring role, the cockpit of the long range passenger carrying aircraft will be used for examples and argument. The cockpit has always been at the forefront of technology. The configuration of the cockpit is driven by all the issues bearing on the man-machine interface, and because it is so insulated from the outside world, the cockpit must contain, either within it, or connected to it by communication systems, all the solutions to the problems of long duration equipment monitoring. The modern cockpit is where many human activities have been automated and the role of cockpit personnel changed from procedural specialists to an equipment monitor and man-machine interface manager. The advent of ultra-long haul flight will further distance humans from procedural activities. This distancing, or peripheralisation, is intimately related to the degree, as well as the style, of automation, and is a potent negative factor in monitoring efficacy.

The problem of long term monitoring in aircraft is not a new one. Pilots have always been aware of the responsibility, the necessity in times of stress, fatigue or conflict within or outside of the cockpit, of first scanning all the instruments and maintaining awareness of the state of the aircraft. Even those involved in the first long distance flights were aware of the disparity between the demands of the monitoring task and the limitations of human monitoring ability. Charles Lindberg, in his crossing of the Atlantic, was aware of the importance of human arousal and the consequences of hypoarousal and monitoring failure. For 34 hours, Lindberg flew without cockpit windows, sacrificing their streamlining effect, some aircraft speed and pilot comfort in exchange for improved pilot performance from direct contact with the external environment (Lindberg, 1953). Nearly 60 years later the Boeing Commercial Airplane Company has provided an alerting system in their latest widebody aircraft, the Boeing 747-400. Again, the disparity between the demands of the monitoring task in long distance commercial flight and human monitoring ability have resulted in aircraft modification, with its associated certification costs and probable changes in human behaviour. The effectiveness in combating hypoarousal of both Lindberg's open cockpit and Boeing's alerting system is unknown. Neither device monitors the state of alertness of the crew or provides the means whereby the crew might alter their vigilance level.

Despite the recent immense changes in technology and the increasing shift to predominantly monitoring tasks, there has not been an associated development in vigilance assisting systems. Over the last 50 years there have been sporadic attempts to develop such systems, but there has been little support. The doyens of vigilance research dismiss human alerting systems as impractical, cumbersome, expensive and ineffective. Thus, we are now in the position of expecting many people to monitor, yet we have not created a work environment where those monitoring can do so effectively.

It is proposed that the workrole of those who monitor systems upon which considerable human and financial resources depend is becoming one where personal fulfilment and work satisfaction are being progressively ignored and eroded. The persistent inability to meet an unobtainable performance specification is a significant factor in job dissatisfaction, itself a potent factor in degrading the ability to sustain attention. Thus, a key consequence of automation, with its attendant distancing or peripheralisation of humans, is change in human performance which itself interferes with the monitoring process.

It has to be accepted that there is no single theory which adequately describes the behaviour of humans who have been peripheralised into the roles of system monitor and back-up system. Unfortunately, it is probable that there is not enough time for mankind to wait until human attentional systems are understood sufficiently such that either human monitoring can be manipulated or that job design can be altered to fit human abilities. Immediate measures must be taken to reduce the present mismatch between the demand for monitoring services and the inadequacy of the supply, even though these measures must be crude and inefficient. Peripheralisation with monitoring failure is likely to have been a factor in the Chernobyl nuclear reactor accident. A repeat of this monitoring failure is something that mankind cannot afford.

Part A
Monitoring Problems and Processes

2 Automation, peripheralisation and error

The modern aircraft cockpit allows humans to find safe and cost effective solutions to the problem of transporting large loads over long distances through a hostile three dimensional maze. In this tiny workplace a highly motivated workforce interacts with a very advanced and rapidly changing technology. This new technology is essential on economic grounds, but much of this technology is not accessible to those on the flight deck, humans being distanced, or peripheralised, from essential flight processes (Norman, Billings, Nagel, Palmer, Wiener, Woods, 1988). Part of this peripheralisation process may stem from aircraft design failing to focus on human needs (Wiener and Curry, 1980). In this section, automation induced peripheralisation and its effects are considered as are modulating factors. After a brief examination of aircraft accident types, the interrelationships between automation, peripheralisation, human error and present responses to human error are considered. The consequences for monitoring in the cockpit are discussed.

2.1 Automation in aircraft

The external appearance of commercial aircraft has altered dramatically as their efficiency as a transport system has increased. Propulsion units have changed from low power, poorly reliable, fuel hungry piston engines to high thrust, ultra reliable, fuel efficient turbofans. Automation, defined as the replacing of human function with machine function, has been and will remain a key factor in obtaining optimal performance from modern propulsion units. Modern aerodynamascists have mated this propulsion efficiency to airframe developments which allow very large aircraft to carry big loads at relatively high mach numbers over prodigious distances. Again, automation has been a key factor, a variety of automated airframe functions now being essential for keeping costs at levels which allow airlines to supply seats at competitive rates. Another less visible technological revolution, as radical and as essential as the others, has been the automation of aircraft control systems. Automation of aircraft control systems has been essential for realising the economic potential of propulsion and aerodynamic developments. The prevailing philosophy underlying automation of the human interface of aircraft control systems has been questioned (Norman et al., 1988; Wiener, 1988; Rogers, 1991), partly because human factors accidents now predominate, the same human factors issues appearing repeatedly (Learmount, 1992). New, more human-centred approaches such as adaptive automation have been proposed (Morrison, Gluckman and Deaton, 1991; Emerson and Reising, 1991). These new approaches, and their potential contribution to cockpit monitoring management, are considered later.

2.1.1 Automation and aircraft control systems

Modern flight control systems are almost completely automated. In some instances humans are actively discouraged from using them because the control system requirements are too demanding (Billings, 1989). There are many examples, some well established like the yaw damper, and many, like neutral stability flight systems, automated flight controllers and automated aircraft warning systems, which are currently in service at various stages of development.

The need for automated flight control systems is obvious. Swept wing aircraft yaw away from banked turns. Such aircraft are fitted with an automatic device called a yaw damper which counteracts this tendency. This device is not usually turned off because human correction can be tedious and demanding (Davies, 1979; Hawkins, 1987). Similarly, humans have difficulty flying aircraft with neutral stability. There are significant reductions in drag and hence decreases in fuel costs if aircraft are flown with neutral stability, but computer assistance is essential for flying aircraft in this economically desirable configuration (Hopkins, 1987). In the same way, the flight path which best satisfies safety and commercial demands is best determined by an automated flight controller. These are but a few of a myriad of examples where human procedural, information processing, and management capabilities are being exceeded. Humans are becoming the limiting factor in systems performance (Taylor, 1989) in many respects. The introduction of automated warning systems is further evidence of the help humans need in the management of the modern commercial flight deck. Thus, there are aspects of flight control which are now impractical for humans, but without which commercial aviation becomes uneconomic (Tsang and Vidulich, 1989). Automation in the modern aircraft cockpit is widespread and essential.

2.1.2 Cockpit crew duties - old and new

Before the advent of modern control systems, piloting required considerable physical strength under certain flight conditions (Davies, 1979). In other circumstances, marginal levels of thrust could demand a very delicate touch. Cockpit crews would manually check fuel levels at stopovers, keep fingers crossed about tyre temperatures and guess about wind shear. Some crew workload was present during the cruise phase of flight, because autopilots were relatively primitive, navigation did not have the benefit of inertial navigation systems and flight deck instrumentation only gave direct information about aircraft systems. Crew workload increased in the vicinity of airports because of unsophisticated air traffic control systems, increased aircraft density and the manual demands of landing. Crew workloads were relatively high during most phases of flight.

The tasks and duties of cockpit personnel in the latest commercial aircraft are totally different. Every phase of flight can now be carried out without human intervention, and for economic reasons, many aspects of flight management are best performed by automated flight control systems. During the cruise phase, there is only a monitoring role as flying controls, engine management, fuel management, cabin environment control and navigation are all automated. There are few direct links between cockpit displays and critical aircraft components. Many of the monitoring systems only display if there is a problem, a multitude of systems sharing the same final display unit within the cockpit. Some aircraft do not have direct links between the cockpit controls and the control surfaces, the aircraft being managed by parallel computer systems which examine and evaluate the human input and dismiss it if it would cause danger to the aircraft (Hopkins, 1987). Crew workload in cruise is limited to monitoring, the workload at times being extremely low. Some aircraft can fly so far that machine endurance far exceeds the maximum acceptable duty period of pilots, some airlines using one crew for cruise, and the other for take-off and landing. In the vicinity of airports workload can fluctuate. While air traffic control has become very much more sophisticated, it has been swamped by the recent large increases in air traffic. Flight path changes during descent and approach can increase workload dramatically and unpredictably (Hughes, 1989; Wiener, 1989).

2.2 Peripheralisation

The term 'peripheralisation' describes the process of role change which accompanies increased levels of automation (Norman et al., 1988). Peripheralisation is a complex psychobiological state which occurs as a consequence of automation. It has been proposed that automation results in role changes where humans are shifted from being in direct contact to being a machine prosthesis, a system maintainer or a system manager. The automation of flight control systems, such that they cannot practically be accessed by humans, has resulted in significant role changes for flight deck personnel (Wiener and Curry, 1980; Roscoe, 1992), peripheralisation being an inevitable consequence.

2.3 Peripheralisation effects

Peripheralisation occurs in the cockpit, but its effect on humans is ill understood and little considered. Peripheralisation effects can probably be managed in many situations, providing there is an appropriate balance during the process of automating the man-machine interface between being 'human-activity centred' and being 'task-requirement centred'. This balance is unlikely to have been achieved in the cockpit. Several have suggested that the human-centred component has been lacking (Wiener and Curry, 1980; Norman et al., 1988). Peripheralisation in the cockpit produces a variety of effects believed to be important in human error production. These effects, complacency, miscommunication, and changes in situational awareness shall be considered in turn.

Peripheralisation is likely to be a dynamic process (Degani and Wiener, 1991). Thus, under optimal conditions the functional interface between man and machine may be close to that intended. This interface is an aggregate of the human-hardware interface, the human-software interface and, as recently suggested, a human-procedural interface

(Degani and Wiener, 1991). Under certain circumstances there will be a shift in the position of one or more of these interaction areas, and the overall functional interface may end up far from that intended. Illustrative examples include the input of false information into flight computers to make automated systems behave as desired (Wiener, 1989) and the changes in interface position which occur consequent upon the automation of checklists (Palmer and Degani, 1991).

2.3.1 Complacency

Complacency is a behavioural category used in incident classification in the National Aeronautics and Space Administration's (NASA) Aviation Safety Reporting System. Complacency has been defined in this context as 'self-satisfaction which may result in non-vigilance based on an unjustified assumption of satisfactory system state' (Parasuraman, Bahri, Molloy and Singh, 1991). While this definition is helpful, particularly in incident analysis, the extent to which complacency is a problem induced by automation can be difficult to delineate from other subtle changes in crew coordination and behaviour (Norman et al., 1988). However, recent studies suggest that there is a significant linkage between automation and complacency, key factors being the consistency and reliability of the automated system (Parasuraman et al., 1991). While complacency and a decline in situational awareness can occur simultaneously, or the former can precede the latter, decline in situational awareness can occur without any apparent evidence of complacency. There are a variety of human behaviours considered in this section on complacency, even though they are significantly different in a number of ways. These behaviours all have elements of non-vigilance coupled with assumptions about the state of control systems.

Primary/secondary task inversion Human behaviour is rarely unaffected by changes in machine prowess. 'Primary/secondary task inversion' (Wiener and Curry, 1980; Palmer and Degani, 1991) is the behavioural phenomenon where the presence of a backup system, such as an altitude alert mechanism, results in flight crew using the backup system as a primary source of information about altitude. Similarly, the best result from using various checklist systems occurs with the traditional challenge-response method, inferior results being obtained with variably automated systems, such as a manual-sensed checklist or an automatic-sensed checklist (Palmer and Degani, 1991).

Automation deficit Although clearly different from the above examples, automation deficit, that is the temporary and relative reduction in performance on resuming a task which has been automated, has an element of complacency in it. It may be possible that this type of complacency can be managed by adaptive automation (Ballas, Heitmeyer and Perez, 1991).

Boredom-panic syndrome Human peripheralisation has been linked to a behavioural response, the boredom-panic syndrome. During the boredom phase the automatic system is able to cope, but when its abilities are exceeded, a human operator may panic upon being suddenly thrown into a highly active and sometimes crucial role. According to pilots with varying Boeing 757 experience, there is too much programming required below 10,000 feet in the terminal area. This sudden workload after extended monitoring can have negative effects on crew roles (Hughes, 1989). The boredom-panic syndrome is particularly prevalent where there are extremely negative consequences linked to relatively rare events such as in nuclear power plants (Norman et al., 1988). The boredom-panic syndrome is an extreme behavioural effect of peripheralisation, but is not limited to man-machine interfaces which are being progressively automated. Thus, a commonly held view of the anaesthetists' workrole is 'hours of boredom punctuated by seconds of stark terror'. In this example, automation has not resulted in peripheralisation, but complacency and a lack of situational awareness are key elements. Boredom-panic usually coexists with complacency, although complacency does not always progress to this more extreme behavioural response. This example serves as a reminder of how interrelated these complex human behaviours can be. It is likely that human error potential is increased during the panic phase of the boredom-panic syndrome.

Complacency is an insidious effect of peripheralisation. It is likely to remain a problem as long as the philosophy underlying automation of the man-machine interface is more task-centred than human-centred.

2.3.2 Communication

Effective crew coordination and resource management are associated with characteristic information transfer processes. Effective crews have different methods of communication compared with ineffective crews, the former being characterised by a frequent, direct, open and concise communication style (Kanki, Lozito and Foushee, 1989; Kanki, Greaud and Irwin, 1989). The ephemeral nature of human communication suggests that it should be vulnerable to peripheralisation (Wiener, Chidester, Kanki, Palmer, Curry and Gregorich, 1991).

There are many examples in human factors accidents which suggest that there is a relationship between peripheralisation and ineffective communication. It has been suggested that there is a relative decline in amount and quality of crew communication in advanced technology aircraft, which supports previous anecdotal evidence of subtle automation induced changes in crew coordination (Norman et al., 1988). In a comparison across three Boeing aircraft types with varying degrees of cockpit automation, interpilot communication declined as the level of automation increased (Costley, Johnson and Lawson, 1989). In many of the recent incidents and accidents involving faultless aircraft with highly automated flight decks (3.4.2), communication has often been inappropriate for the flight situation (Habsheim, 1990; Hill, 1990). The separate contributions from complacency and poor communication cannot be easily discriminated, but the presence of a peripheralisation effect is difficult to refute.

Part of the communication problem might relate to the sheer quantity of information that each pilot must share in order to keep the whole crew informed. Each new display must be incorporated and encoded by the crew (Taylor, 1989), it being vital that a proportion of the information be shared. In a survey of late model Boeing aircraft pilots, there was concern with the management of the quantity and display of information, a proportion of pilots wanting a 'declutter' function that would sift out what was vital (Chandra, Bussolari and Hansman, 1989). Information glut appears to be a problem for pilots (Mosier-O'Neill, 1989), much more data being aggregated and particularised for effective flying and decision making (Maher, 1989).

Thus, it is probable that peripheralisation impairs human communication, particularly in the context of man-machine interfaces which are more task-centred than human-centred. Multiple factors may modulate the effect of peripheralisation on communication, both positively and negatively (2.5).

2.3.3 Situational awareness

In a commercial context situational awareness is the accurate perception of the factors and conditions that affect an aircraft and its flight crew during a defined period of time (Schwartz, 1989). Other definitions have added comprehension to perception, and have also included the projection of their status in time (Tenney, Adams, Pew, Huggins and Rogers, 1992). Situational awareness in a military context is the skill of maintaining an awareness of the tactical situation. At present the air force views it as the single most important factor in improving mission effectiveness.

Situational awareness is an aggregate of many perceptual and cognitive processes. Important features of a satisfactory state of situational awareness are the presence of the 'big picture', access to information and an appropriate cognate state.

The 'big picture' The loss of the 'big picture' is a feature of many human factors incidents and accidents. In relatively automated aircraft, automation induced peripheralisation is a potent factor in reducing situational awareness. The Airbus A320 accidents at Habsheim, Bangalore and the recent accident on approach to Strasbourg (3.4.2) demonstrate the tragic consequences of a flight crew losing almost all situational awareness (Habsheim, 1990; Hill, 1990; Learmount, 1992). In less automated aircraft, loss of the 'big picture' has also occurred all too easily. A much quoted incident involved a China Airlines 747 near San Francisco. Crew preoccupation with an engine malfunction at high altitude, and the loss of awareness of airspeed decline, resulted in a stall and spin in which the structural limits of the aircraft were exceeded (Norman et al., 1988; Tenney et al., 1992).

Information acquisition Peripheralisation from automation can produce a decline in situational awareness because system design has stopped pilots obtaining information essential for predicting system status in the near future. Thrust reverser control systems have been modified after the Lauda Air 767-300ER accident in which the deployment of one thrust reverser in flight preceded the loss of the aircraft (Norris, 1991). The Lauda Air crew appeared powerless to manage the uncalled for reverser deployment, the automated engine control systems not readily providing information that was actionable. In a similar way, the implementation of datalink for routine air traffic control, currently part of the Federal Aviation Authority National Airspace Plan, has caused concern because of the loss of 'party line' information (Hansman, Hahn and Midkiff, 1991). Situational awareness in the final approach phase may be impaired, because the highly automated communication system could exclude adventitious information from other aircrews.

Cognate state The availability of information as a factor in situational awareness has to be distinguished from the appropriateness of the cognate state for integrating the information. At present, the next generation of widebody aircraft are offering an electronic library system, where modern computing power will allow flight crew unfettered access to information of all classes. It is not clear that the electronic library system is associated with systems which might optimise the cognate state of those requesting more information. Again, it appears that the processes for providing appropriate information on flight decks has a bias to being task-centred rather than human-centred.

Summary Situational awareness is vulnerable to automation induced peripheralisation. The processes underlying peripheralisation directly threaten situational awareness, the availability of information and appropriateness of the cognate state being amplifying factors. Situational awareness changes, out of all the peripheralisation effects, are probably the most lethal.

2.4 Human error

The consequences of peripheralisation all appear to increase the likelihood of human error in an automated surrounding. It is worth looking at the causes of aircraft incidents and accidents to see how significant human versus machine errors are and what types of human error occur.

2.4.1 Aircraft safety

There is no doubt that flying has become safer. In 1921 the United States Air Force suffered major mishaps at a rate of 467 mishaps per 100,000 flying hours. At that loss rate the Air Force of today would

Cockpit Monitoring and Alerting Systems

damage or destroy its entire present inventory of 9,000 aircraft in about 7 months. By the 1980s the major mishap rate had dropped to less than two per 100,000 flying hours (Diehl, 1989). The rate for civilian aircraft has dropped from 130 in 1938 to 7.5 in 1986, major commercial airlines now having mishap rates a lot less than one. Differences in accident rates between major commercial air carriers and other civilian aircraft are less dramatic when accident data is expressed in terms of the number of departures (Koonce, 1989).

2.4.2 Error types and human factors

Aircraft accident investigators, not unexpectedly, used to search initially for mechanical problems because of the unreliability of power plants and control systems in older generation aircraft. The label 'pilot error' was

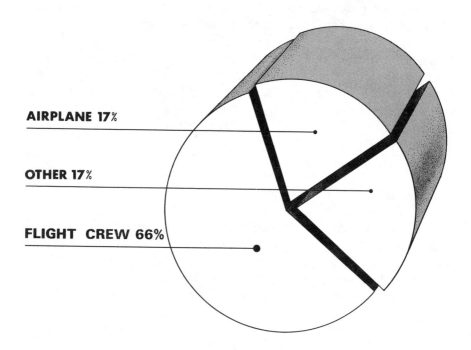

Figure 2.1 Primary cause factors in 738 aircraft hull loss accidents, 1959-1989

Source: Data from Logan and Braune, 1991

used reluctantly for a variety of reasons including investigator training, liability issues and the difficulty of gathering human performance data (Diehl, 1989). A number of incidents and accidents in the last 15 years have highlighted the futility of ignoring human factors. The investigation of the human factors aspect of aircraft accidents is now being systematised (Feggeter, 1991).

In 1975 the United States Federal Aviation Administration (FAA) started a confidential, voluntary, non-punitive safety reporting program, NASA acting as an objective, disinterested third party for the collection of data. Examination of 35,000 incident reports suggested that several categories of aircraft accident involving operational and human factors were subsets of incidents containing the same elements (Billings and Reynard, 1984). Thus, the prevalence of incidents is useful in identifying the frequency and type of human related problems in aviation in general, and in the cockpit in particular. In 22,226 consecutive reports, aircraft problems constituted about 4 per cent while flight crew error constituted about 49 per cent of incidents (Billings and Reynard, 1984). The FAA study and others have suggested that human error causes or has a contribution to 50 to 90 per cent of all aircraft accidents (Yanowitch, 1977: Feggeter, 1982; Billings and Reynard, 1984; Hawkins, 1987). With respect to commercial jet aircraft accidents, the Boeing Commercial Airplane Group has demonstrated that the flight crew were the primary factor in accidents with known causes (Lautman and Gallimore, 1989; Orlady, 1990; Logan and Braune, 1991). In the period from 1959 to 1989 flight crew were the primary factor in 66 per cent (Figure 2.1), with little change in the period from 1980 to 1989 (Logan and Braune, 1991).

The leading factor (33 per cent) attributable to crew was deviation from basic operating procedures (Degani and Wiener, 1991). The FAA study suggested that many of the incidents revealed multiple factors. Thus, failure of information transfer was present in 70 per cent of incidents, primarily related to voice communication. However, information transfer problems were rarely the sole factor. Many reported human errors were specifically behavioural in origin, psychological factors being far more important than perceptual or medical factors. Psychological factors cited in the FAA study included complacency, motivation, attitudes, workload and distraction (Billings and Reynard, 1984). Resource utilisation problems were attributed to training deficiencies, although psychological factors may have had a large part to play, further increasing errors of a behavioural origin.

Many errors are likely to have faulty crew dynamics as a factor, errors being at a team level (Brown, Boff and Swierenga, 1991). The difference in communication patterns in relatively mistake prone simulator crews (Kanki, Lozito and Foushee, 1989) suggests that crew dynamics can produce negative consequences with an enhanced error potential (Brown et al., 1991).

The aggregation of key factors in incident analysis obscures many findings which are specific for incidents and accidents of certain types. In military helicopter error analysis, poor information exchange or poor management of workload occurred in 75 per cent of instances (Leedom, 1991). Analysis, using a common error taxonomy, of military, airline and general aviation accidents revealed that decisional errors were similar in all classes (approximately 50 per cent) but there were marked variations in procedural and perceptuamotor errors (Diehl, 1991).

2.5 Peripheralisation error modulators

The consequences of peripheralisation are complacency and changes in crew communication and situational awareness (2.3). It is likely that all these effects are significant factors in incidents and accidents, as many of them have been observed in situations where faultless aircraft have come to grief (2.3.1-2.3.3). The process whereby automation induced peripheralisation produces effects which promote human error is complex and like all error producing processes, whether peripheralisation related or not, is open to modulation (Figure 2.2). It is important to consider all possible modulators[1], despite the uncertainty about underlying mechanisms, because of the possibility of identifying an ameliorating factor. There is little point in attempting to rank the strength of either peripheralisation effects or modulating factors, because ranking requires objective measurements which are generally not available (2.6.1).

Error modulators can be classified in a number of ways, division into fixed and variable groups implying that the latter may be relatively malleable. Some modulators are not clearly in either group, and these have been collected together in an 'other' category. Variable modulators

---

[1] The term 'modulator' is used instead of the term 'moderator' to avoid confusion with the concept of moderator as applied to stressors (4.1.1).

are crew structure, complacency, stress, workload, sleep cycle and fatigue. Fixed modulator factors include leadership, personality, organisational culture and age. Other factors include training and crew selection amongst many.

2.5.1 Variable factors

Crew structure Commercial cockpit crews are not permanent. Most cockpit crews are together for a duty cycle, some never flying again as a team in a large airline. Learning curve effects can be dramatic at the

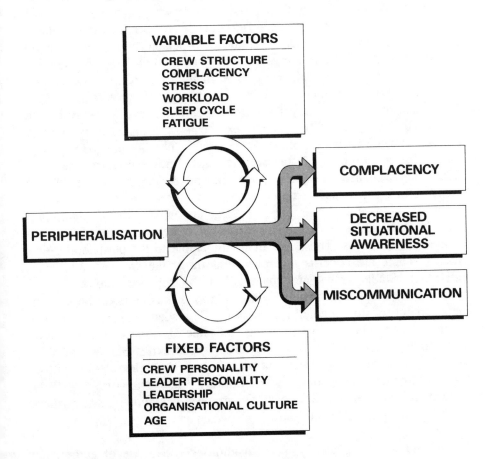

Figure 2.2 Peripheralisation produces effects which are open to modulation by a variety of factors

start of a duty cycle (Chidester and Foushee, 1989). However, airlines have persisted in creating new crews for each duty cycle, despite evidence which seriously questions this approach (Foushee and Helmreich, 1988). Thus, the team experience is a variable which affects human performance, being able to act as an inhibitor or amplifier of error producing processes. The military uses a stable crew structure, as well as team management training, in trying to optimise crew situational awareness. Analysis of the relationship between communication and performance suggests that commercial crews that have recently flown together make fewer errors and communicate more efficiently than crews that have not recently flown together (Brown et al., 1991).

Significant crew performance differences have been observed related to crew size. In an extensive study, two-person crews consistently outperformed three-person crews in both line and LOFT operations, this difference being independent of the level of cockpit technology (Clothier, 1991). The mechanism behind the modulator effect of crew size is unknown, although effects from group cohesiveness may be the reason. The crew size effect is complex because expansion of the human resource appears to be beneficial once trouble has occurred (Predmore, 1991). In a number of major incidents, crew performance was outstanding because crew size was expanded, the need for human resources being matched to the demand.

The stability and nature of crew structure appear to be significant error modulators, specifically in areas of crew communication and situational awareness. These are areas vulnerable to the peripheralisation process. There are benefits in crews thinking of the crew as a separate individual. Stable crews are characterised by this separate individual evolving some specific mental skills, such as metacognition, the explicit management of team cognitive activity (Thorsden and Klein, 1991). Stable groups, both during training and line operation, appear to be essential for the evolution of higher order group skills. Stable crews are more likely to be able to carry out planning which is flexible and allows for contingencies (Leedom, 1991; Conley, Cano and Bryant, 1991), rather than new crews which meet performance demands by standardised but relatively inflexible activities.

Complacency Complacency is an insidious consequence of automation and peripheralisation. It is likely that it is a modulating factor in many error producing processes, independent of whether it has been produced by automation induced peripheralisation or not. It has been considered previously in some detail (2.3.1).

Stress There are a variety of stresses that have been considered in cockpit crews (4.1.1, 4.3.1-4.3.2), such as life stress, the stress of certification and the stress created by non-routine events, called acute reactive stress. Other sources of stress have been proposed (Thomas, 1989; 1991). Stress produced as a consequence of automation induced peripheralisation has been relatively ignored. It is likely that there are specific types of stress associated with monitoring (Hancock and Warm, 1989; Hancock, 1991), which occur in some individuals within crews and more rarely in whole crews (Warren, Hudy and Gratzinger, 1991). There is considerable conceptual and factual confusion surrounding stress and stress management. These concepts and their implications for cockpit management are considered later (4.1-4.2).

Anecdotal evidence suggests that stress is a potent modulator of error propensity. There is little doubt that error production in cockpits has frequently occurred on a background of life stress, but it is unknown if all types of stress have the same ability to modulate error production. It has been proposed that increased stress results in a restriction of cue utilisation (5.2.1), which has consequences for situational awareness. In inexperienced individuals, stress produces simplification of recalled information, further reducing situational awareness (Beringer, 1989).

While it appears that stress is likely to promote peripheralisation induced errors, some have suggested that stress has performance enhancing effects. There are terminological and conceptual problems with the entity of useful stress (4.2, 4.2.2). While public performers and athletes commonly attempt, and appear to succeed, in using some of the performance enhancing effects of stress, or more correctly increased arousal (4.2), there is no evidence that cockpit crews use such techniques, nor is it certain that routine use is effective. Stress and its implications for performance are considered elsewhere (5.2.2, 5.2.4).

Workload Intensity of workload has been cited in a variety of instances as a factor in reduced crew performance (Hughes, 1989). Workload is likely to be a significant modulator of peripheralisation effects, over a third of Boeing 757 pilots disagreeing that automation reduced workload (Wiener, 1989). Others have suggested that this proportion is much less (James, McClumpha, Green, Wilson and Belyavin, 1991). There are a number of incidents where the increased workload in the complex environment of some major airports, coupled with the perceived necessity to program the automated control systems, rather than 'click them off' (Wiener, 1989), has led to cockpit crews losing their situational awareness in a very significant way.

There are levels of workload that might well be too low for optimal performance. When workload levels are low, the potential for complacency and boredom should be inversely related to workload. There is anecdotal evidence to support this. It is possible that the error potential is not related to the level of workload, but more to the rate of change of workload. Automation deficit, the reduction in performance on resuming a previously automated task (Ballas et al., 1991), is a situation where the change in workload might be the critical factor. Altering workload appears to be a primary goal of some cockpit design groups. Workload measurement and its role in cockpit design and cockpit alerting systems is considered elsewhere (7.7.1).

Sleep cycle Changes in performance caused by disruption of sleeping patterns is a topical cockpit issue (Graeber, 1986; 1988; Gander and Graeber, 1987; Dinges and Graeber, 1989) and of obvious importance to crew with a primary monitoring function in ultra-long haul flying. The importance of the sleep cycle as a modulator of error production is not clear (O'Hare and Roscoe, 1990). The benefits from programmed napping on fatigue levels and performance (Dinges and Graeber, 1989) suggest that sleep cycle effects are significant. The relationship between sleep deprivation and complacency, situational awareness and communication is likely to be complex. The sleep cycle and its effects on arousal and monitoring are considered later (5.4.2-5.4.3).

Fatigue Fatigue has many similarities to workload with respect to modulating the peripheralisation process, but it captures cumulative effects. Fatigue is important because it is a marker, poorly defined, of the appropriateness of the cognate state for integrating information. At face value, it should be a factor in situational awareness.

Simulator studies using complicated and demanding schedules have suggested that crews which have been together, but were fatigued, performed better than rested off duty crews (Kanki, 1991). Whether this result is transferable to line operations is unclear, although some studies suggest that it can. One study on line operation incidents has suggested that the number of incidents declines as the duty cycle progresses and does not increase as the number of flight hours increase (Logan and Braune, 1991). In these predominantly domestic operations, more than half of the incidents occurred in the first hour, suggesting that if fatigue is to be considered seriously as a factor it must be viewed in a global and cumulative way. The time course for the cumulative effect is unknown but is likely to vary both within and between individuals.

Flight crew have relatively entrenched views on the detrimental effects of fatigue and the role of scheduling in fatigue production. These views probably stem from the much greater incidence of incidents and accidents at the end of flights than at the beginning. The greater number of incidents at the end of flights can be due to a multitude of factors, the evidence cited above suggesting that fatigue is not a major cause. It is likely that there are levels of fatigue where the potential for enhancing peripheralisation effects are balanced by usage, familiarity, crew dynamics and efficient communication.

2.5.2 Fixed factors

Leader personality has preoccupied those interested in cockpit human factors (Chidester and Foushee, 1989; Kanki, Palmer and Veinott, 1991). This preoccupation has obscured the importance of a number of separate issues, such as cockpit leadership, and the personality of other aircrew.

Crew personality The importance of personality in cockpit crews is one of the few topics that significantly polarises human factors practitioners. There have been many instances in aircraft incidents and accidents which suggest that the personality of individual crew members is a factor in error propensity. This type of observation must be treated with caution as there is a tendency to explain behaviour in terms of personality traits rather than as a result of situational forces (Brown et al., 1991). In the formation of crews, individual personality may be a factor in altering error potential, particularly in the face of peripheralisation effects like complacency and change in communication.

Leader personality Cluster analysis of a variety of personality attributes with the assessed quality of leadership confirm a relationship between personality and cockpit leadership (Kanki, 1991). Accident proneness has not been accepted as a stable personality characteristic (Dolgin and

Gibb, 1989), but there are personality types that are not suited to cockpits (Sellards, 1989) because of their ability to impair communication and coordination (Helmreich and Wilhelm, 1989).

Leadership personality has been predictive with respect to crew performance in simulator studies, having a major effect when crews first come together, while information transfer, a behaviourally based variable, is the dominant factor with time (Mosier, 1991). Isolated examples, such as the failure of all engines in a Boeing 747 from the ingestion of volcanic dust, and the saga of United flight 232 (Predmore, 1991) suggest that leadership personality is a factor in crew performance in line operations, particularly in difficult circumstances. It is unclear if leadership personality can modulate peripheralisation induced errors.

Leadership Cockpit leadership has been considered under general leadership constructs (Blake and Mouton, 1985; Gibson and Wilhelm, 1989) but there are factors which distinguish the cockpit from all other workplace situations. The consequences of sub-optimal flying during unexpected events can be lethal to all cockpit members, suggesting that there is a threshold below which mutual concern for task or person is irrelevant (Jensen and Biegelski, 1989). The survival factor suggests that the leadership role may be dependent upon specific behaviours. Simulator research supports leader behaviour as being more important than leader personality in predicting crew performance (Mosier, 1991). The flying personnel of United Airlines have emphasised that decisiveness is an indispensable element in the leadership process (Blake and Mouton, 1985). The continuous examination by each cockpit member of crew performance suggests that the earning of idiosyncratic credits in an exchange theory approach to leadership may be more appropriate to the cockpit situation. Leadership which detects and proclaims the presence of peripheralised behaviours may be a modulating factor of the peripheralisation process.

Organisational culture There have been a number of serious non-aviation accidents where organisational factors were key precursors (Reason, 1991; Johnston, 1991). Organisational characteristics have been proposed as a significant factor in a series of aircraft accidents (Reason, 1991; Gunn, 1991; Kahn, 1991) including the Air Florida Boeing 737 at Washington National Airport in 1982 (3.4.1), the USAIR Boeing 737 at LaGuardia in 1989 and the Air Ontario Focker F28 at Dryden in 1989.

Organisational factors have been collected together under the umbrella term of 'organisational culture'. Organisational culture reflects management style, values and behaviour and is sensitive to organisational change induced by disputes, mergers and deregulation (Hayward and Alston, 1991; Gunn, 1991). Organisational culture also reflects the culture in which it is embedded, and this may be a relatively ignored but important factor in flight deck management (Helmreich, 1991). More egalitarian organisations have less barriers between flight crew and other sections, promoting safety. Access to information is a factor in situational awareness, and a culture that reduces barriers and promotes a free interchange between various divisions is likely to partly demystify some of the peripheralising aspects of automated systems.

In this discussion, organisational culture has been classified as a fixed factor. This is obviously not true in a negative sense in that organisational change often damages a culture with respect to the safety of operations, but is true in a positive sense in that many cultural change programs struggle hard to demonstrate real effects. Management practitioners are constantly promoting ways in which to change organisational culture, and some management styles are particularly attuned to removing barriers and promoting synergies (Moss Kanter, 1989). The effectiveness of these approaches in some organisations has been demonstrated, but the effectiveness for flight crews is unknown.

Age Age is a factor in modulating the effects of peripheralisation, but is complex. Younger flight crew cope with computer systems, flight deck programming and changes in technology with relative ease (Wiener, 1989). In contrast, older crew are less perturbed by the decline in manual flying which occurs in more automated aircraft. Age as a factor may be overstated (Mortimer, 1991), as effects may reflect changes in education, values, societal-cultural issues and workrole expectations.

2.5.3 Other

A variety of factors, which might be modulators of peripheralisation effects, could be discussed. These include training, crew selection, drugs, gender, culture and new technology. New technology is considered separately (2.6), because new technology has been seen as a solution to peripheralisation effects. Only training will be considered further.

Cockpit Monitoring and Alerting Systems

Training It is unknown whether training reduces peripheralisation effects, but it is likely that a deficient training program provides a background on which peripheralisation effects can flourish.

A common complaint of trained and experienced pilots flying the Boeing 757, which has a moderately automated cockpit, is ignorance related to the 'design intent' of automated systems (Orlady, 1989; Wiener, 1989). Situational awareness thrives on incomplete information, but there must be limits to the amount of information that can be incorporated into a training program. The on board electronic library is seen as a partial antidote to the incomplete information problem. Some have suggested that poor communication style, which can be promoted by peripheralisation, should be susceptible to training.

While increased transfer of information in the training stage is not a realistic solution to peripheralisation, there is little doubt that inadequate training promotes peripheralisation. In a number of accidents in which situational awareness was a factor, there were crew training deficiencies.

2.6 Responses to errors

Over the last decade, aircraft incidents and accidents have frequently resulted in immediate alterations to aircraft control systems and structures. These changes have been an essential factor in improving air safety. Often an incident or accident has resulted in aircraft systems undergoing further automation. Some have suggested that manufacturers and aviation authorities have responded to new incidents and error forms with a technology rather than a human-centred approach (Norman et al., 1988). The continuing manner in which automation is being applied raises questions about the priorities driving the design of the man-machine interface. The problems surrounding a technology-centred approach are being dramatically demonstrated at this present moment.

Air Inter, whose A320 crashed during descent into Strasbourg early in 1992, has now advanced its plans to fit a ground-proximity warning system (GPWS), having originally had a pilot's head-up display as a higher priority in its cockpit fit (Learmount, 1992). The three approach and descent accidents in the A320 (3.4.2) may be unrelated to the recommendation of the Direction Générale de L'Aviation Civile that GPWS be compulsory, although this is unlikely. The extent of the automation within the cockpit of the A320, and its purported infallibility, may be a factor in the reluctance to incorporate additional automated systems. The addition of another automated system is unlikely to be the real solution, as in all A320 accidents there has been ample information about the threat to the aircraft. The two previous A320 accidents, where faultless aircraft landed unintentionally causing total hull losses and fatalities (Hill, 1990), resulted in software changes to the engine autothrottle controller (Moxon, 1991). Further changes to automated systems are likely after the Strasbourg accident.

Other examples of the technological responses to human error include recommendations for changes to engine controls after the Kegworth accident and the move to use datalink air traffic control systems because of the multiple instances of inaccurate information transfer. Human error, incidents and accidents often result in more automation.

2.6.1 A scheme for peripheralisation

Automation and peripheralisation are associated with human error. It is possible to develop a scheme which displays the interactions between automation, peripheralisation and human error (Figure 2.3). A key feature is the circular relationship between these entities once human error results in more automation (2.6). There are positive rotational influences at each stage, economic demands being the initiating factor.

The presence of multiple sites where one factor promotes another, raises the possibility of the automation, peripheralisation, human error cycle revolving without economic demands, once sufficient errors occur. Removal of humans from this scheme would stop the cycle completely. The case for pilotless cockpits is considered below (2.6.3).

Peripheralisation results in human errors via a number of effects which have been discussed previously (2.3.1-2.3.3). While these effects and modulator factor effects cannot be ranked, the ways in which the peripheralisation effects coexist is important. The elements that interconnect peripheralisation and human error could be arranged in a serial or parallel manner, or a mixture of the two. There has been no suggestion that a change in communication always precedes a decrease in situational awareness or vice versa in cockpit incidents. While both are often present at the time of an incident, cockpit transcripts give no clear indication that one precedes and causes another. Similarly, complacency often precedes but also occurs simultaneously with a decrease in situational awareness, but it is not always so. Thus, it is

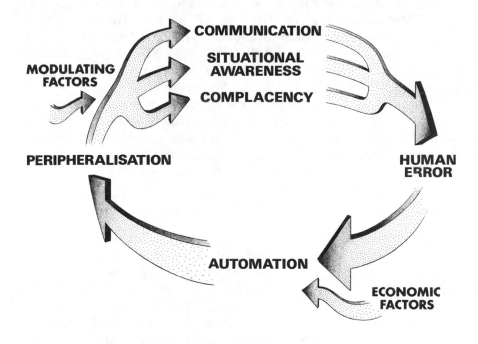

Figure 2.3 A scheme for the interaction between automation, peripheralisation, its effects and human error. For diagrammatic simplicity, the different types of modulating factor are not shown (2.5)

much more likely that there is a matrix of relationships between complacency, communication and situational awareness, and that a serial arrangement where one leads to the other is unlikely.

Many modulating factors are present and almost all of them (2.5) can either promote or inhibit peripheralisation. Some factors appear equally able to promote or inhibit, there being a continuum of effect but others, like lack of training and acute reactive stress, seem more likely to promote peripheralisation rather than inhibit it. Most modulating factors are likely to work at multiple sites.

2.6.2 Future trends

The unwanted consequence of implementing new technology into an incompletely automated environment is peripheralisation related human error. This is unlikely to decrease. Unchecked, and with humans remaining in some increasingly peripheralised role, the opportunity for human error will expand. One response to the error forms that have occurred with cockpit automation has been the development of CRM programs. The effectiveness of this response is considered later (3.3).

2.6.3 Role for humans

The technology required to totally remove humans from cockpits has been present for at least a decade (Norman et al., 1988). Although commercial aircraft rarely fly free from any human input, the Trident, now disappearing from service, has had full autoland capacity for over a decade. Modern military aircraft can carry out a full mission, including threat management and weapons delivery, without human input. Many of the latest model civilian aircraft could carry out pilotless commercial operations with relatively little modification.

The arguments against removing humans have been considered (Wiener and Curry, 1980; Norman et al., 1988). In commercial aviation, passengers are resistant to the concept of pilot-free flight. Other arguments relate to the unforeseen occurrence which human laterality and creativity will always manage better than computer software (Pew, 1986). A classic example is the fuel exhaustion incident of an Air Canada Boeing 767, where the pilot, whose private hobby was gliding, used his hobby skills to land the widebody twin on a glider strip with no power. Similarly, space shuttle personnel have made unplanned space walks which have ensured successful deployment of expensive satellites. The sagas of United Flight 232 and many others remain a testimony to human skill, creativity and judgement (Predmore, 1991). There is little doubt that there will continue to be a role for humans in the monitoring of modern technology systems.

2.7 Peripheralisation and monitoring

Monitoring requires sustained attention or vigilance. Vigilance is a human activity that is relatively vulnerable, despite its essential contribution to species success. Vigilance has an inherent capacity to decrease and can be affected by a myriad of factors (5.1). It is not surprising that vigilance should be vulnerable to effects such as complacency and reduced situational awareness.

The ability of peripheralisation effects to damage vigilance is at times staggering. Some cockpit transcripts (3.4.2) suggest that cockpit crew in faultless aircraft have been provided with clear dynamic displays of a rapidly deteriorating state in the situation of their aircraft, and yet these displays have not even engendered a comment. Of even greater concern is the incidence of vigilance related mistakes in excellent crews in automated cockpits where the robustness and tolerance of aircraft and air traffic control systems have ameliorated harmful effects. The recent move to non-punitive incident analysis within airlines has been very revealing with respect to vigilance related errors in automated cockpits.

2.8 Summary

Automation has provided a myriad of benefits. The underlying drive for further automation is an economic one but the cost effectiveness of many modern control systems resides in a specification which is outside human ability. A natural corollary of this type of automation is the peripheralisation of humans from a controlling role. A variety of unwanted changes in man can be attributed directly and indirectly to the peripheralising process. There must come a time when the costs of these consequences will offset the economic benefits of automation. These costs are unpredictable, particularly as the consequences have the potential for societal costs which are very difficult to measure. Peripheralisation cannot be allowed to continue unchecked. Humans will retain a role in control systems, albeit in monitoring, because of their unique abilities, but the prospects for them contributing significantly will decline as long as the present goals of the automation process remain unchanged.

3 CRM as a response to peripheralisation

Peripheralisation has multiple effects. Complacency and changes in communication and situational awareness are factors in the promotion of human error, human factors incidents and accidents (2.6.1). A variety of human factors accidents have been cited as part of the motivation for the development of cockpit resource management (CRM) programs. In this section, the origins, objectives, key concepts and effectiveness of CRM programs are considered. Some human factors accidents are reconsidered in the light of peripheralisation effects and their possible modulation by CRM. The ineffectiveness of CRM in providing a complete solution to human factors accidents is discussed.

3.1 CRM origins

Ten accidents, occurring between 1972 and 1982, have been cited as the motivation for creating CRM programs (Jensen and Biegelski, 1989). Many CRM programs have used one or more of the ten accidents in their introduction or discussion (Margerison, Davies and McCann, 1986; Maher, 1989). In addition, conferences hosted by NASA as well as the famous NASA based Ruffel Smith study were also significant factors in the initiation of CRM (Houston, 1990). NASA researchers noted that aircrew used a wide range of management behaviours in coping with a specific task. This observation suggested that business management concepts, which promoted uniform management practices, might profitably be applied to cockpit operations (Povenmire, Rockway, Bunecke and Patton, 1989; Houston, 1990). Management analysis of the ten accidents suggested that captains failed to make effective decisions because they, or their crew, did not use effectively all the available resources. There is little doubt that resource issues are still a significant problem. In some recent human factors accidents automation may be changing the problem of inefficient resource utilisation. Human factors accidents are reconsidered later (3.4.1-3.4.2).

3.2 CRM objectives and key concepts

The style, content and the theoretical basis of CRM programs have changed since their introduction in 1979. The goals of the first programs were to reduce the number of incidents and accidents which were behavioural in origin. The emphasis was on changing behaviour of individuals, rather than of crews, using instructors to alter the individual's knowledge base. Evaluation of program effectiveness was not a feature of the first programs.

The latest programs view ineffective group processes as being the underlying issue. Skills-based training is preferred to activity-based training. Programs limited to enlarging the individual's knowledge base are outdated, as are instructors presenting information. An evaluation plan is a preferred feature, as are process monitoring systems which facilitate continuous change (Freeman and Simon, 1991). Skills-based training uses facilitators who focus on processes, often drawing on information and behaviours that are already present. At present, there are relatively few programs which use facilitators and are skills-based (Komich, 1991).

Early programs attempted to educate individuals in a number of concept areas. Some of these areas, such as communications, crew coordination and decision making, still constitute core material. Other original areas included personality, fatigue, and stress. This latter group is less amenable to skills-based approaches. With time there has been refinement in the focus of CRM courses. Research has shown that attitudes in communication and coordination, command responsibility and recognition of stressors are related to crew performance (Helmreich, Foushee, Benson and Russini, 1986; Gregorich, Helmreich and Wilhelm, 1990). In contemporary programs, themes of crew communication, coordination, and decision making are viewed from a small group perspective (Brown et al., 1991). Programs are enhanced with additional themes such as recognition of stressors and the understanding of situational awareness.

There has been a growing realisation that behavioural change does not necessarily follow a raising of awareness of crew coordination and the changing of attitudes. A number of additional factors can interfere with

CRM as a Response to Peripheralisation

behavioural modification (Swezey, Llaneras, Prince and Salas, 1991). Factors include variation in the initial skills base of participants and the necessity for reinforcement of behavioural change by feedback and practice (Swezey et al., 1991). Cultural factors, both inside and outside the organisation, are likely to be important in behaviour modification (2.5.2).

At present there is significant diversity in the style and content of CRM programs. Some are advocating shifting the focus from the cockpit to the whole crew and renaming cockpit resource management as crew resource management. Unfortunately, collaboration in the production of CRM courses has been limited. Some sharing of program material has occurred, but this is uncommon and there appears to be a great deal of unnecessary duplication. Diversity in CRM program style and content should reflect cultural (organisational and other) differences, but cultural differentiation does not explain the present variation between programs. Economic and marketing issues appear to be significant factors.

Most CRM programs are still similar to the initial programs in style and content, there being some activity-based components, but skills-based approaches are rare. Program content will be briefly reviewed, as this dictates the status of crews flying at present.

3.2.1 Decision making

Decision making skill enhancement has a traditional two pronged approach in many contemporary programs. Discrimination ability is promoted by cognitive training techniques which use acronym driven models to organise decision making processes, while attitude training consists of recognition of hazardous patterns and incorporation of appropriate antidotes. Acronyms have included DECIDE (detect, estimate, choose, identify, do, evaluate), PASS (problem identification, acquire information, survey strategy, select strategy) and others. Activity-based training is carried out in the classroom using a variety of technologies, the simulator, line orientated flight training (LOFT) exercises, and actual aircraft. Teaching effectiveness has been demonstrated in the short term, both in civilian (Jensen and Bielgelski, 1989) and military operations (Povenmire et al., 1989). The two pronged approach which encompasses rational and attitudinal components has been modified in pilot judgment training by recognising individual styles and optimising their interactions with circumstances (Westerlund, 1991).

3.2.2 Communication

Communication effectiveness is usually examined as a facet of management style analysis, virtually all CRM programs using some form of relationship versus task model. Cockpit communication is a central focus of CRM training because it is the means whereby cockpit resources are managed. Aspects of communication which are desired in aircrew are inquiry, advocacy, listening, conflict resolution and critique. Failure in one or more of these aspects has contributed significantly to error production (2.3.2). While the communication patterns of effective crews have been characterised and the effects of automation noted (2.3.2), there are changes in communication with workload which suggest that team-based styles, which have an implicit component, are important (Bowers, Morgan and Salas, 1991). Activity-based courses focus on underlying interpersonal sensitivity and openness, but they may not be able to generate the efficient abbreviated styles required for high workload situations. Communication efficiency in CRM programs is evaluated by written test, peer feedback, observer feedback, and video tape. Self assessment is most commonly employed. Conflict resolution groups and crew discussion after LOFT exercises may be effective in modifying communication style in the short term. Most current training programs still focus on individual skills rather than on team communication strategies (Swezey et al., 1991).

3.2.3 Leadership, followership, team concept

An essential part of CRM programs is leadership training. This includes both external and self assessment via a variety of techniques including LOFT exercises. Just as important is the training of the 'follower' in management and decision making. The development of maturity and an understanding of leadership is required of cockpit members apart from captains. The ability to be a follower is required of captains when first officers are in command. These training goals are intimately linked with the team concept and the power of synergistic activity. There have been implementation problems, because the prevailing method of personal assessment is still on an individual basis. While synergism remains an important goal in CRM programs, delegation is also essential, leadership and followership roles being intimately intertwined with synergism and delegation (Jensen and Bielgelski, 1989). Leadership, followership and a team concept have been reinforced in more recent activity-based courses by focusing on attitudes underlying the sharing of responsibility.

3.2.4 Others

A proportion of contemporary programs mention stress, even if it is only to acquaint flight crew with relevant types of stress. Some programs have employed parts of generic stress management programs (4.4.2). The best method for incorporating stressor recognition, or the recognition of human fallibility under stress, into activity- and skills-based programs is unknown. Activity-based programs, which expose aircrew to combinations of incidents which are far removed from the most pessimistic of scenarios, reveal a conceptual confusion. There is no good evidence that activities which attempt to produce a state of stress which mimics acute reactive stress are effective in teaching stressor recognition. These types of activity-based programs mix the issues of workload management, acute reactive stress and stressor recognition. If this mixing is unfocused, concepts and strategies are imperfectly acquired. The ideal activity-based program achieves a coordinated, group-based behavioural style related to workload distribution, cross monitoring of performance and the recognition of strain in partners.

Similarly there are behavioural modification schemes (Pawlik, Simon and Dunn, 1991) which may develop situational awareness skills in a training situation. Their effectiveness is open to question as the skills attained may be consistent with the aims of the program, but different to the complex, shared team skills which are required for effective crew situational awareness.

3.3 CRM effectiveness

The effectiveness of CRM programs is currently being examined. The first wave of CRM programs exposed many pilots to CRM training, but virtually all these programs were begun without an evaluation system. CRM programs can be assessed in a variety of ways.

3.3.1 Incident and accident analysis

Some have asserted that the relative rarity of commercial aircraft accidents has been the reason for not using them in a prospective manner for assessing efficacy of CRM programs (Helmreich, 1991; Clothier, 1991). While accident rates are extremely low the rate of incidents is not. If it is assumed that accidents are a subset of a much greater pool of incidents containing the same elements (Billings and Reynard, 1984), then a retrospective study of human factors incidents in which crew have or have not been CRM trained should have already demonstrated the usefulness of CRM programs. In time, a prospective study using such an approach should be able to test whether a skills-based approach is superior to a knowledge-based approach.

Incident analysis is possible and useful, judging from the taxonomy of 244 human error incidents produced from 410 safety investigations in one airline's operations (Freeman and Simon, 1991). Other airlines appear to be using this technique as a means of monitoring and altering training. It is unclear why this approach has not been used for testing the effectiveness of CRM programs. Costs may be an issue, although the present methods of course assessment are not cheap. Control of sensitive information may be the major hurdle, although it is possible to sanitise this sort of data.

The prior exposure of cockpit personnel to CRM programs does not figure prominently in incident and accident reports for reasons unknown. There have been exceptions. For instance, neither of the crew of a Boeing 737-400, which carried out a rejected take-off at La Guardia's Runway 31 on September 20, 1989 killing 57 passengers, were CRM trained (Pope, 1991).

There has been a tendency to attribute a CRM trained crew's good performance in coping with an incident as a positive effect from exposure to the CRM program. United, Continental and USAir have all done this (Jensen and Bielgelski, 1989). This is unacceptable evidence (Helmreich, Chidester, Foushee, Gregorich and Wilhelm, 1990) and CRM advocates have been belittling their cause by citing single supporting case reports.

It is difficult to accept that the continuing high incidence of human factors accidents is only occurring in crews who have not had CRM training. In the first six months of 1992, 11 out of 16 (69 per cent) accidents were provisionally placed in the human factors category (Learmount, 1992), similar to the incidence of five or ten years ago. The extent of CRM training in the major airlines and the fact that many regional airlines have also completed first generation CRM programs

CRM as a Response to Peripheralisation

(Wilson, 1989) together suggest that CRM training may be less of an antidote to human factors accidents than originally hoped. This is not to say that CRM programs have no point. There is the distinct possibility that CRM programs modulate some but not all the processes involved in human factors incidents and accidents.

3.3.2 Program evaluation

There have been evaluation approaches which have focused on the behavioural changes produced by CRM programs rather than on reductions in crew error. Initial attempts failed to use methodologies which could appreciate behavioural change. A preliminary study on B52 bomber crews suggested that CRM training enhanced combat mission performance (Povenmire et al., 1989). Individual airlines have been certain that there has been a change in attitudes to crew interactions (Taggart and Butler, 1989; Wilson, 1989).

Although empirical evidence for CRM program efficacy has been difficult to obtain (Helmreich and Wilhelm, 1989), one study has examined the effectiveness of programs using course assessments, rather than whether aircrew involved in incidents were CRM trained (Helmreich et al., 1990). Comparing attitudes prior to training with those following training revealed significant increases in measures such as command responsibility, recognition of stressor effects and communication and coordination. It has been suggested that these short term benefits were to be expected (Helmreich and Wilhelm, 1989). The measurement of CRM program effectiveness has progressed. There has been an extensive assessment of flight crew performance embodied in the NASA/University of Texas crew performance project (Helmreich, 1991). A collaboration between airlines, NASA, FAA and the University of Texas has provided a cross-fleet and a cross-airline facility for monitoring CRM programs and line operations.

Check airman and LOFT instructors have assessed crew behaviour in airlines using a checklist derived from the Cockpit Management Attitude Questionnaire (Helmreich, 1991). Behaviour measures were assessed with Likert intensity scales (Clothier, 1991), the measures being combined to assess attitudes in three categories: command responsibility, recognition of stressor effects and communication and coordination. These categories have been validated as predictors of human performance (Helmreich et al., 1986). CRM training significantly improved a variety of measures, the improvement being most marked with the initial program though still present, but smaller, with recurrent training (Clothier, 1991; Irwin, 1991). The improvement was not uniform across all measures, being largest for briefing and self-critique and least for task concern and technical proficiency. Assessment on either the line or LOFT showed significant training effects (Clothier, 1991). The categories in which attitudes were captured were not changed uniformly by CRM programs, 20 to 30 per cent of pilots showing a negative attitude change in one of the three categories, 3 per cent reacting negatively on all subscales. There was a significant decline in attitudes in all three categories during the intervals between training interventions, the most pronounced being in the communication and coordination area (Irwin, 1991).

It is safe to conclude that CRM programs have definite short term effects on altering attitudes and behaviours, although the effects are not uniform. The long term consequences are less clear, some form of recurrent training being necessary. Recurrent training is costly and is relatively less effective. Some have suggested that achieving greater behavioural changes will require an immense investment, there being a distinct possibility that there will only be marginal improvements (Stragisher, 1991).

In summary, it is still not clear that there is a relationship between the advent of CRM programs and a decline in the frequency of human factors incidents and accidents. There is the possibility that CRM training does not produce behavioural changes useful in managing peripheralisation effects, and that these behavioural changes are not even measured in the present assessment approach. Indeed, a new scale measuring attitudes regarding automation and the operation of automated systems is under evaluation at present (Helmreich, 1991).

3.4 CRM, accident processes and peripheralisation

In the accidents which played a part in the evolution of cockpit resource management programs (Jensen and Bielgelski, 1989), there usually was an initiating problem (Table 3.1). Problems with fault lights, ATC messages, engines and icing were unremarkable, part of routine cockpit operations and repeatedly covered in training and certification programs. However, these problems were associated with mismanagement of other

Table 3.1
Human factors accidents, 1972-1982

Airline Site, Date	Aircraft	Features
Eastern Miami, 1972	L-1011	Undercarriage extension problem. Situational awareness deficit about flight path. Descended into swamp.
PanAm/KLM Tenerife, 1977	B747 / B747	ATC message confusion. Situational awareness deficit about take-off conditions. Runway collision.
United Portland, 1978	DC8	Landing gear fault light problem. Situational awareness deficit about fuel state. Fuel starvation.
National Escambia Bay, 1978	B727	Failure to receive start descent point. Ignorance about descent profile plus high workload. Uncontrolled approach.
Air Florida Washington, 1982	B737	Wing icing at take-off. Situational awareness deficit and multiple cockpit factors (3.4.1). Failed to fly.

cockpit functions, automation induced peripheralisation having a role but not a dominant one.

CRM training should be useful in the management of cockpit resources in situations where there are multiple factors which promote error production. These situations continue to cause accidents, some recent human factors accidents having much in common with those of a decade ago (Table 3.2). Some problem related human factors accidents have been very complex and a challenge for any CRM program.

3.4.1 Problem related human factors accidents

Two problem related human factors accidents are described. In both accidents there were error producing processes which CRM training

Cockpit Monitoring and Alerting Systems

Table 3.2
Human factors accidents, 1983 - 1991

Airline Site, Date	Aircraft	Features
Air India New Dehli, 1986	B747	Horizon control reversal on FDI. Situational awareness deficit. Flew into water in a banked dive after take-off.
Delta Dallas/Ft Worth, 1988	B727	Incorrect flap deployment, incomplete pre-flight checks. Failed to take-off.
BMA East Midlands, 1989	B737	Engine failure at end of ascent. Good engine shut down. Situation awareness deficit. Crashed short of field.
USAIR La Guardia 1989	B737	Take-off problem. Rejected take-off mismanaged. Poor understanding of cockpit functions.
Air Ontario Dryden, 1989	F28	Wing icing at take-off. Situational awareness deficit and multiple cockpit factors (3.4.1). Failed to fly.

addresses. There was little evidence in either accident of automation induced peripheralisation.

Air Florida Flight 90 In this Boeing 737, which crashed on taking off from Washington DC National Airport in 1982, there were repeated muted expressions of concern by the first officer concerning ice accretion and take-off performance (Margerison et al., 1986; Maher, 1989). The failure of advocacy on the first officer's part and the failure of listening on the captain's part have been much discussed. The major external factor that this aircrew should have been able to manage was airframe icing. This accident revealed cockpit management problems with communication and coordination, judgment and decision making and command responsibility. Air Florida was a recently formed airline

CRM as a Response to Peripheralisation

and many of the training programs of other airlines were not in place. Both crew members were relatively inexperienced, particularly in those specific circumstances. Effects such as complacency and poor situational awareness might have been present, but if so, they were not due to automation induced peripheralisation. Rather, training problems, with deficiencies in available information, and possibly in the cognate state, were dominant factors.

Air Ontario Flight 363 In this Focker F28, which crashed on taking off from Dryden Airport in 1989, both crew knew that snow was falling onto the wing while waiting for take-off. The option of de-icing was not available because an unserviceable auxiliary power unit and the unavailability of ground power made it impossible to stop the engines, a necessary prerequisite before de-icing. Two pilots, who were passengers, expressed concern to cabin crew about the amount of ice on the wing, but the crew did not contact the cockpit. Stopping the engines would have stopped the flight, an unattractive option as Dryden did not have adequate accommodation for the passengers. The take-off decision was also affected by the need to make onward connections for passengers, the recent merger involving Air Ontario, as well as the next sector being the last of a duty cycle for the crew. The accident revealed problems with judgment, decision making and communication. Poor situational awareness was present but not due to automation induced peripheralisation.

3.4.2 'Faultless' aircraft in human factors accidents

Other accidents have occurred in the last ten years in which there has been no coexistent fault or problem. The presence of automation induced peripheralisation is difficult to dispute. In these accidents there was little opportunity for a CRM program to have provided an effective antidote.

Avianca Flight 11 In this accident in 1983, the aircraft was faultless and no adverse external factors were present. The Boeing 747 landed short near the outer marker with heavy loss of life, there being a number of contributing human factors, discussed in detail by others (O'Hare and Roscoe, 1990). Situational awareness was significantly impaired, a chart reading error, which resulted in the aircraft being well below the correct altitude, being aided and abetted by the crew ignoring the GPWS.

Cockpit Monitoring and Alerting Systems

Attempts to reset the inertial navigation system (INS) in the final phases of flight were not appropriate. The poor communication both within and outside the cockpit, the reduced situational awareness and the generalised complacency, as evidenced by the multitude of small errors made over the last thirty minutes of flight, together suggested that significant peripheralisation had occurred. The resetting of the INS, the lack of response to the GPWS, and the failure to monitor each other or the aircraft suggest that automation itself may have been the culprit.

Air France A320 This highly automated airliner, only two days in commercial service, was flown over the small general aviation airport of Habsheim in 1988 for air display purposes. Both pilots held management positions, one being the airline's A320 technical pilot, but neither had flown in an air show, the presence of passengers being unusual in this situation. At no time was cockpit communication or command responsibility satisfactory in the fly past. The altitude of the aircraft was never stabilised before the plane crashed into trees at the end of the runway and was destroyed. The aircraft was faultless in all respects (Hill, 1990). In the context of perfect flying conditions and the absence of external factors, this A320 accident was dominated by reduced situational awareness, impoverished communication and possibly significant complacency. These effects, in the face of the automated protection systems in this type of aircraft, strongly suggest automation induced peripheralisation was a major factor.

Indian Airlines A320 In a visual descent into Bangalore in excellent flying conditions, this three month old aircraft landed 700 metres short, being totally destroyed with heavy loss of life. The aircraft was faultless. A check captain and a trainee captain entered inappropriate descent settings into various parts of the automated cockpit systems, both acknowledging the wrong settings but neither monitoring the descent (Hill, 1990). Communication was inappropriate and command responsibility was deficient. Reduced communication and reduced situational awareness was associated with significant complacency. Peripheralisation, produced by cockpit automation, was probably enhanced by a lack of understanding of the nuances of the automated systems.

Air Inter A320 Initial evidence suggests that there was an inappropriate input into the flight control unit such that the aircraft approached Strasbourg in a relatively steep 3,300 ft/min descent rather than the much gentler, normal descent of 700-800 ft/min (3.3°). The communication pattern in the cockpit before the descent started was inefficient and marked by a number of errors. The cockpit voice recorder revealed no concern over the rapid descent and the large changes in instrument displays (Learmount, 1992), suggesting that situational awareness was deficient, as occurred in the other A320 accidents. Again, situational awareness decline, complacency and altered communication were present. Automation induced peripheralisation appears to have been a dominant factor.

3.4.3 Summary

There is evidence in some recent human factors accidents of reduced situational awareness, inadequate communication and significant complacency. In these accidents there was no evidence of external factors or minor fault conditions. While manageable internal faults or external factors were and still are a feature of many human factors aircraft accidents, it is possible that there are now different processes underlying error formation at the man-machine interface.

3.5 Automation, peripheralisation and CRM

Airlines attribute human factors accidents to flight deck mismanagement. A counter proposition is that a proportion of human factors accidents are now produced by automation induced peripheralisation and no amount of cockpit management training in its present form will reduce the frequency of human errors. Cockpit management training, the airlines' response to human factors accidents, may be a tangential solution for a proportion of the errors in automated cockpits.

Attempts at altering flight deck management must not be belittled, because uniform behaviours are beneficial, while error modulators (2.5) are almost certainly changed by some aspects of CRM programs. Further, CRM programs are being altered into culturally sensitive, skills-based change programs. The alternative views of CRM program efficacy are best considered schematically (2.6.1).

Cockpit Monitoring and Alerting Systems

3.5.1 Optimistic view

CRM programs could alter automation effects if they were able to influence the cycling of the automation, peripheralisation, human error loop (2.6.1). To stop automation effects, CRM programs must break the circular path completely, the most suitable area being the stage between peripheralisation and human error where a number of modulating factors can be changed (2.5). The daunting prospect for those holding the optimistic view is that CRM programs must simultaneously sever all links between peripheralisation and human error. If peripheralisation is still able to effect either complacency, situational awareness or communication, human error remains likely (Figure 3.1).

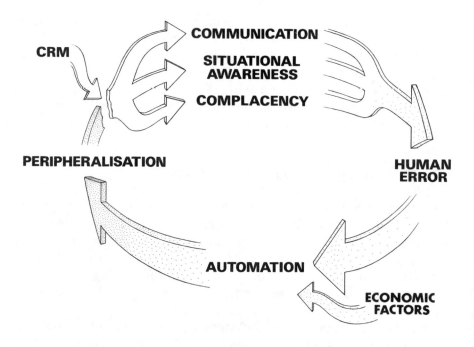

Figure 3.1 An optimistic view of the effectiveness of CRM programs on the interaction between automation, peripheralisation and human error. See Figure 2.3 and 2.6.1 for details

3.5.2 Pessimistic view

CRM programs probably improve cockpit communication, although their effects may be short lived (3.3.2). The effectiveness of CRM programs in altering situational awareness and complacency must be doubted. Situational awareness deficits are obvious in recent human factors incidents and accidents, despite most pilots having being exposed to CRM programs in some form (3.3.1, 3.4.2). A pessimistic view would have that CRM programs only influence the inadequate communication path that exists between peripheralisation and human error. Other paths, their malignant positive influences unaffected by CRM, can keep the automation, peripheralisation, human error loop cycling (Figure 3.2). At

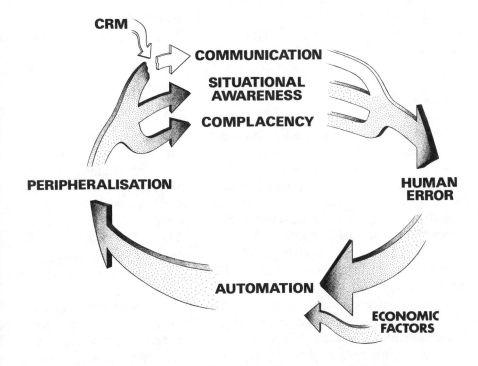

Figure 3.2 A pessimistic view of the effectiveness of CRM programs on peripheralisation effects. CRM may optimise communication but the other factors remain capable of producing errors

present, human factors practitioners are attempting to measure communication (Kanki, 1991). Until the measurement of situational awareness (Tenney et al., 1992) and complacency is similarly developed, there is little prospect of usefully altering CRM programs.

3.5.3 A measured view

The effectiveness of CRM programs in reducing human errors lies somewhere between the two extremes (3.5.1-3.5.2). This assessment will be unsatisfactory for proponents of CRM programs. However, unless all involved in CRM programs can maintain an impartial view as to program intent and effectiveness, the persistence occurrence of human factors incidents and accidents may obscure the real benefits that CRM programs can provide. The use of incident analysis in controlling and refining the processes underlying CRM programs is the correct approach.

3.6 Summary

CRM programs were created to combat an apparent new error form, the human factors accident. It is unlikely that human factors accidents are a recent phenomenon. Rather they have become more obvious as other causes of accidents have been abolished. CRM programs were initiated because different parts of the aircraft control system, be it crew, software or hardware, were contributing information, but it was being ignored or managed inefficiently. The solution, improved cockpit communication and decision making, has provided some immediate benefits. Long term benefits and accident prevention are not proven sequelae.

Human factors accidents are likely to result from heterogeneous processes. Some of these accidents are dominated by complacency, miscommunication and changes in situational awareness. Other accidents show some evidence of these peripheralisation effects but they are not as prominent. CRM programs may be doomed if viewed as the complete solution to human factors accidents, because CRM programs may address issues which are tangential to the basic problem, the peripheralisation of crew by cockpit automation. At present CRM can do little about automation induced peripheralisation.

4 Stress and arousal in cockpits

Stress plays a large part in human environmental interactions. Humans do not change as rapidly as technology changes, making the incorporation of new technology potentially costly to humans in a variety of ways. This book is concerned with human behavioural responses to automation, with particular focus on the behavioural responses associated with the monitoring role. Terms like 'stress' and 'arousal' will be used frequently, but these terms have been used in many ways without precise or consistent definition (Sanders, 1983; Levine, 1986). Stress has already been considered briefly as an error modulator in the peripheralisation process (2.5.1). In this section, stress and arousal are defined and a start is made on how the processes underlying stress should be considered. This background is used to discuss the known causes of stress in cockpits, before the identification and consideration of the new stressors relevant to peripheralisation and the monitoring role.

4.1 Stress

Confusion arises with the term 'stress' because of the widespread roles that stress is given in the workplace and in disease production. These roles are of particular concern for groups with vested interests, including pilots, but terminology varies significantly depending upon context. A brief description of the traditional view of the stress mechanism is given, before other ways of understanding this entity are outlined and stress is defined.

Stress is not a new entity. Cannon (1914) used the term stress to describe emotional states that had possible detrimental physical effects on organisms. Selye (1956) proposed that there was an entity called the General Adaptation Syndrome which was the body's response to prolonged debilitating circumstances of many different types. The response was non-specific with respect to the inciting agent and had elements of alarm, resistance and exhaustion which resulted in stereotyped patterns of hormonal secretion. The patterned pituitary-adrenocortical stress responses were produced by an undefined primary mediator. Selye's studies were not primarily occupational or workplace related but were more physiological. Research on occupational stress using survey and laboratory methods has demonstrated significant effects on organisational and individual performance (Kahn, Wolfe, Quinn, Snoek and Rosenthal, 1964), as well as having significant health consequences (Quick, Horn and Quick, 1986).

Selye's view of stress is a static one, each individual having a different ability to resist the eroding effects of some types of life activity, these abilities being whittled away in different ways and at different rates. In this static scheme, stress can be satisfactorily defined as the interaction between a force, action or situation that places demands upon a person and the person's psychological, behavioural and physiological responses. In understanding how Seyle's static view of stress interacts with the peripheralisation process, it is necessary to conceptualise the processes underlying stress.

4.1.1 A static stress scheme

There are many models of the processes underlying stress. The basic entities that are commonly present include stressors, moderators, strains and consequences. Stressors, strains and consequences are usually interconnected sequentially with moderators acting at various sites, affecting the interaction between the sequential elements. The static view lacks feedback and has been called an open-loop process. These entities and their relevance to the cockpit will be considered briefly.

Relevant stressors Temperature, noise and vibration are typical environmental stressors for pilots. Apart from these, three others are commonly recognised (4.3.1), namely stress associated with non-routine events or acute reactive stress (Green, 1985), stress associated with proficiency checks (Cooper, 1985), and stress associated with non-work activities or life stress (Alkov, Gaynor and Borowsky, 1985; Sloan and Cooper, 1985; Simmel, Cernovnik and McCarthy, 1989). Others have much larger lists of stress factors (Thomas, 1991), but most ignore the stressors critical to the monitoring role in automated cockpits (4.3.2).

While there appears to be a large number of stressors, it must be remembered that the presence of a stressor does not necessarily result in stress or an alteration in performance. The presence of significant moderators will obtund stressor action, reducing the amount of strain and making stress related consequences less visible.

Cockpit moderators Moderators are individual characteristics which modify stressors (Matteson and Ivancevich, 1987). There are many modifiers, some of the accepted ones including the tolerance for ambiguity (Kahn et al., 1964), Type A behaviour (Ganster, 1986), locus of control (Marino and White, 1985), negative affectivity (Brief, Burke, George, Robinson and Webster, 1988) and self-focused attention (Frone and McFarlin, 1989). With respect to the cockpit, it has been suggested that the incentive to do well (O'Hare and Roscoe, 1990), which encompasses a number of the accepted modifiers, moderates most stressors.

Cockpit strains Strains or outcomes are behavioural, psychological and physiological. Common behavioural strains include smoking, reduced productivity, increasing alcohol and drug abuse, accident proneness, irritability, violence, job dissatisfaction and absenteeism (Quick et al., 1986). Common psychological strains include reduced morale, poor motivation, family problems, sleep disturbance, sexual dysfunction, depression and burnout. Physiological strains include changes in blood pressure and heart rate (Kaufmann and Beehr, 1986; Frankenhaeuser, Lundberg, Fredrikson, Melin, Tuotmisto and Mytsten, 1989), increases in adrenalin, cortisol and cholesterol as well as changes in skin conductance (Bruning and Frew, 1987; Fusilier, Ganster and Mayes, 1987). Most studies concerning stress, the health of pilots, and their fitness to fly use a static view.

Consequences The consequences of strain can be measured by job satisfaction, organisational commitment, illness production and turnover intention. Illness production is easily accommodated in Selye's view of stress mechanisms. None of these, nor illness production, are presently relevant to modern commercial pilots. The turnover intention of pilots is relatively low and the job satisfaction relatively high in common with other work roles dominated by procedural activities.

Illness rates during the working life of pilots is low, although these low disease rates do not translate into a relative longevity. Health in retirement is not relevant to the issues being considered. If it is accepted that disorders like coronary artery disease and hypertension have a stress component in their aetiology, then it appears that all these stressors are either relatively mild or there have been powerful moderators at work, because the stress profile of pilots is different from that of other professions (Simmel et al., 1989). In the Mayo Clinic study on pilot health (Holt, Taylor and Carter, 1985), the cockpit personnel age adjusted coronary artery rate of 2.6/1,000 person-years is considerably less than the benchmark Framingham data (>8.0/1,000). Even allowing for selection bias, this lack of stress related illness contrasts strongly with the disease patterns of those in other monitoring roles, such as air traffic controllers, whose coronary artery disease rate is four times the community average. Thus, cockpit personnel have not shown the normal consequences of stress. Possible ameliorating factors may be the time course of stress, the external origin of the stressors and the motivation of pilots. O'Hare and Roscoe (1990) suggest that in the short term all these stressors are unimportant in aviation, simply because they can be overcome by a sufficient incentive to do well, a property that these authors believe pilots have in abundance. The presence of new stressors (4.3.2) complicates this view.

Summary There are benefits from using the static scheme, even though its open-loop approach misrepresents underlying mechanisms. The principal benefit is simplicity and the opportunity to avoid the consideration of specific central nervous system mechanisms. It is unlikely that this reductionist view can be justified in the context of stress, arousal and human behaviour in cockpits.

4.1.2 Other stress schemes

In many ways the static view of stress is unsatisfactory. It is unable to come to terms with 'useful stress', and it ignores the complex, dynamic, feedback-driven behaviour patterns which give humans the ability to buffer both transient and semi-permanent causes of disequilibrium (Levine, 1986).

Another way of viewing stress in man uses a system approach. The essential feature of the system approach is that an organism's internal stability results from the integrated use of continuously and dynamically organised responses. There is no reserve to be used up, there is no 'amount' of resistance that becomes progressively eroded. Rather, there are a variety of integrated behavioural responses, which are flexible in their own right, and which are mediated by complex interactions between many parts of the neural, neuro-endocrine and endocrine systems. This integrated, programmed and organised system view is supported by the earliest stress researchers (Cannon, 1914) and by others who have re-evaluated Selye's static scheme (Mason, 1971). The system view allows for the evolution of neural system interactions and neural networking which are uniquely new with respect to coping with stressors, particularly novel ones. In contrast with the static view, the system view is characterised by interaction and feedback, much of the stress mechanism being closed-loop rather than open-loop.

The approach adopted in this book uses the system view of stress. The system view allows the integration of a variety of concepts, like arousal, attention and useful stress into a scheme for understanding the processes involved in monitoring. It is still possible to use the entity of the stressor in the system view of stress. Stressors no longer work through unidentified mediators which impinge on endocrine systems. In the system view, stressors drive integrated neural and neuroendocrine networks (5.4) which are likely to be specific to the stressor and probably specific even to the context that the stressor occurs in. Stress is defined after stress and arousal have been distinguished (4.2.1).

4.2 Stress and arousal

The terminology related to 'arousal', like much of that related to stress, has a variety of meanings, many causing confusion. Stress and arousal are even used interchangeably. Lay views of the term arousal are often equated with pleasant events unrelated to behavioural emergencies. There have been some who have seen arousal as an open-loop process, much along the lines of Selye's view of stress, while stress has been seen as a closed-loop process. There has been no tangible benefit from this view. There is no evidence of feedback and non-feedback dominated systems in those parts of the brain which have a part to play in arousal regulation. Another approach is described, which allows pleasant arousal and useful stress to be incorporated.

4.2.1 A continuum approach

The most useful way of discriminating the entities of arousal and stress is to use a system approach and view them as part of a continuum (Levine, 1986). The distinction between the two is that after a perturbation there is a return to a state of pre-activation equilibrium with arousal, but if the systems are unable to return to their original settings, then the system is in a stressed state, or stress is present. Under some conditions of sustained stress, suitable activation can still produce a complete resolution, the pre-activation equilibrium state returning when the accumulated consequences of the state of disequilibrium have been dissipated (Levine, 1986). Thus, arousal is defined as a form of stress where activation is resolved following termination of the stressor or perturbing event. Arousal leads to change in closed-loop system performance, but it is impermanent and does not impair function. Stress is where the closed-loop systems are altered and the original state cannot be attained.

4.2.2 Good stress

Some have considered stress to be physiologically and psychologically useful, but firm evidence is rare and suppositions abound. Adult competence is increased if upbringing is relatively stressed (Glass and Singer, 1972). The psychological advantages of stress include stimulation, enjoyment, challenge and improved productivity (Geare, 1989). Life stress scales have good as well as bad events (Beehr and Franz, 1986) and there has been interest in obtaining an appropriate level of stress in competitive sporting activities.

The entity of 'good stress' or 'eu-stress' is much better handled by using the concept of arousal, where pleasant, non-aversive or motivating stressors produce neural and neuro-endocrine changes which enhance performance, both mental and physical, but which do not produce sustained changes. 'Good stress' may well be a misconception and will not be used further in this book. Thus, arousal can be detrimental or beneficial, but is always reversible. Its beneficial aspects will be reconsidered later. Arousal has been defined as a continuum ranging from coma and deep sleep to the most disorganised behaviour of extreme stress (Malmo, 1959; Duffy, 1972). Arousal can be just as acutely destructive, as any level of stress, both high and low levels of arousal being potentially harmful. Arousal is always reversible.

4.3 Cockpit stressors

Others have considered the various stressors in the cockpit and their comments are useful, whichever scheme is used to understand the stress process (Green, 1985; Cooper, 1985; Alkov et al., 1985; Sloan and Cooper, 1985; Simmel et al., 1989; Thomas, 1989; 1991; Karlins, Koh, and McCully, 1989; Little, Gaffney, Rosen and Bender, 1990). The stressor agents that have been identified will be briefly reviewed, before a new stressor is considered in some detail. Recognition of stressors is one of the attitudes in cockpit crew that predicts performance and the processes underlying stressor action will be considered with a system view where possible.

4.3.1 Existing stressors

Environment Temperature, noise and vibration are environmental stressors that cannot be ignored. These stressors primarily alter arousal rather than produce stress in modern cockpits. While much has been done to make the cockpit environment more like that of an 'office', the relationship between any of these stressors and arousal is not straightforward. Temperature effects on arousal are complex because the neural control systems regulating temperature and arousal levels are interlinked (5.4.2). Some levels of noise and vibration improve arousal and performance together, while a total absence of these stressors is definitely detrimental.

Non-routine events Acute reactive stress, or the stress associated with non-routine events, is the most difficult of all stressors to understand and manage (Green, 1985). Again, non-routine events alter arousal and not stress, and management depends upon optimising the relationship between arousal level and performance (5.2.1-5.2.2, 10.7).

Proficiency checks Proficiency checks are part of a pilots' life, producing stress in some individuals because of irreversible effects (Cooper, 1985). Proficiency checks have profound affects on arousal, well known to those involved in certification. Various methods have been used to dampen arousal fluctuations because the performance of some is reduced very significantly. In these individuals, the arousal related reduction in performance is not evident in line operations. Methods for reducing arousal have included more frequent checks, making each check less of an event, a skills transfer component in the check session, and changing the name 'proficiency check' to 'cyclic'. The effectiveness of these changes has yet to be established.

Life events Stress from non-work activities or life stress (Alkov et al., 1985; Sloan and Cooper, 1985; Simmel et al., 1989; Karlins et al., 1989) is a feature of the pilots' life. Many of these are complex, complicated by a pilot's life style, and are relatively irreversible.

Others There are significant stressors for cockpit crew related to organisational change. Airline mergers, industrial action and the political changes in aviation policy can have a major impact on individuals while remaining outside their control (Little et al., 1990). These are relatively irreversible and result in stress.

4.3.2 New stressors

There are several critical stressors which have been ignored. A classical stressor, role ambiguity, is almost certainly present in the automated cockpit. Role ambiguity is produced by the pilot's interaction with the automation systems on the flight deck. Monitoring itself is likely to be a stressor (Hancock and Warm, 1989; Hancock, 1991).

Role ambiguity There is some evidence that automation produces disquiet for pilots (Wiener, 1989). While doubts about automation are not equivalent to role ambiguity, disquiet about cockpit automation, particularly in relation to the human role in the cockpit, points to the potential for role ambiguity. Thus, pilots flying the Boeing 757 with a relatively automated cockpit give a very mixed response to the automated functions (Wiener, 1989), half agreeing that there are still things that happen that surprise them, a third feeling that they fly the plane as smoothly by hand as by automation, and a third disagreeing that automation reduces the workload. In aircraft from a different manufacturer, automation has not been associated with reduced workload (Wiener et al., 1991). Other studies have produced different results for these test questions, reflecting the experience of the pilots in the sample and their exposure to automated systems (James et al., 1991). Thus, automation has the potential to cause role problems for pilots. There is little doubt that automation has been responsible for role change in cockpits (Wiener and Curry, 1980; Norman et al., 1988).

A variety of studies have examined the mechanisms underlying role conflict and role ambiguity (Kahn et al., 1964; Bedeian and Armenakis, 1981; Kemery, Bedeian, Mossholder and Touliatos, 1985; Howard, Cunningham and Rechnitzer, 1986; Tetrick and LaRocco, 1987; Schaubroeck, Cotton and Jennings, 1989). Antecedent variables such as participation, social support and role overload and their effect on the job tension and job satisfaction induced by role conflict and role ambiguity have been examined. The moderator effects of an individual's characteristics such as type A behaviour, tolerance for ambiguity, neuroticism and others have been studied (Howard et al., 1986). However, transposition of these findings to the role ambiguity created by some aspects of cockpit automation is unwise. There is disagreement in much of this research and its applicability to a particular workrole almost certainly requires study in the workplace. The schemes produced for understanding role ambiguity (Schaubroeck et al., 1989) are based predominantly on the static view of stress and have a limited usefulness in any neuro-behavioural analysis. Nevertheless, role ambiguity is a likely consequence of the progressive automation of cockpit functions and the alteration of role from flyer to cockpit manager. Arousal changes and stress seem inevitable.

Monitoring role The demands of monitoring are likely to alter arousal and may even produce stress. Monitoring depends on vigilance which is a recognised stressor (6.2.2). The necessity in monitoring of maintaining a state of alertness, or vigilance, and the vulnerability and fragility of the relevant neural networks are considered later (5.5).

4.4 Stressor management

There are a variety of measures used to regulate arousal and performance on flight decks, although the distinction between arousal and performance has not been clear in existing approaches.

4.4.1 Recognition of stressors

Some CRM programs specifically incorporate training in recognition of stressors. The goal of these programs is the understanding of human fallibility under stress, the hope being that with efficient communication, workload can be distributed such that each crew member performs in an optimal manner. There is little consideration of arousal and its potential to enhance and inhibit performance, although many of the stressors in the cockpit primarily alter arousal rather than have stress effects (4.3.1-4.3.2). The emphasis in most CRM programs is on avoiding overload and over arousal effects rather than on optimising arousal levels.

If stressor recognition is useful, and there appear to be crew performance changes to support this activity, it is important that relevant stressors are identified. It is likely that some are and some are not. Stressors such as life events, environmental factors, proximity to a proficiency check and the presence of non-routine events are relatively easily identified, and shifting of roles can help arrest the reduction in the coping abilities of the most afflicted cockpit person. This approach fails if a stressor remains unrecognised and therefore undetectable. The inappropriate changes in arousal during monitoring and the presence of role ambiguity are not easily detected.

4.4.2 Stress management

Life stress and the stress of certification are frequently mentioned in CRM programs, both being managed by stress management methods which appear to have short term effects. However, in studies with appropriate experimental designs, these stress management programs have no more immediate effect than unrelated training programs and post training effects are similar in control and trained groups (Murphy, 1986).

4.4.3 Peripheralisation stressors

Peripheralisation has the potential to be associated with the stressor, role ambiguity. Peripheralisation results in increased monitoring, a task which has specific requirements with respect to vigilance. Monitoring is also a cognitive stressor. Role ambiguity is a cognitive stressor and it is likely that cognitive stresses have particular effects on the level of vigilance. The vital interaction between cognitive stressors and vigilance will be reconsidered (6.1-6.5), once the nature of vigilance mechanisms and the theories behind the various views on vigilance processes have been described.

4.5 Summary

Present CRM programs have segments devoted to the recognition of stressors. Some use personal stress management techniques. Neither approach is soundly based with respect to optimising the human resources in the cockpit. The former results in redistribution of workload with no attempt to help the afflicted, while the latter addresses stress issues that are relatively peripheral and do not appear to have effects on pilot health. It is suggested that the continued desire to incorporate stress management in some form indicates that there is a stress problem in the cockpit, which at present has not been identified but which is already disturbing flight deck crew. Peripheralisation is producing new, insidious stressors. Peripheralisation is also producing role changes, the new roles like monitoring being a potent source of hidden stressor activity.

5 Vigilance mechanisms

Automation in the modern cockpit has changed the flying role into a monitoring role. Now humans must give sustained attention to software displays which describe the state of the aircraft as a transport device. Sustained attention, or vigilance, is the ability of observers to maintain their focus of attention and to remain alert to stimuli over prolonged periods of time (Warm, 1984). A persistent and dominating feature of the vigilance process is its instability, which manifests as the vigilance decrement, or the decline in vigilance with time. The various vigilance theories will be briefly reviewed, before considering those that are particularly relevant to the monitoring role in long range commercial aircraft. The vigilance process results from organised patterns of neural activity in nervous system networks that are still being characterised. Our understanding of the vigilance process has resulted from the use of a variety of indirect and direct investigatory methods. The characteristics of the neural networks responsible for vigilance will be considered as will the links with the neural networks important in arousal regulation.

5.1 Vigilance theories

At present there does not appear to be one dominant and universally accepted theory of vigilance, as no one theory is entirely satisfactory for explaining vigilant behaviour (Koelega, Brinkman, Hendriks and Verbaten, 1989; Galinsky, Warm, Dember and Weiler, 1990). In this book, the terms vigilance and sustained attention are considered to be the same, but distinct from the phenomenon of attention. The basic phenomenon of attention involves a selection process (Jennings, 1986), which is the choice of events to be observed and acted upon (Hillyard and Hansen, 1986).

This book draws a distinction between the processes underlying the sustained attentional state and the processes responsible for the onset of the attentional state. This distinction may be artificial in that the vigilance decrement could reflect the natural decline in the activity of the system responsible for attentional selection. However, the nervous system responds differently at the onset of the attentional state and during sustained attention. The focus of interest in this book is the state of sustained attention or vigilance.

The first theories of vigilance tried to explain the vigilance decrement by using the psychological modelling techniques that were current at that particular time. Other theoretical approaches have reflected the current learning, neurological, psychophysical or information processing models of behaviour (Loeb and Alluisi, 1984). These will be considered briefly as each contributes something to our understanding of vigilance.

5.1.1 Learning models

The learning models based on the 'inhibition' of Hullian conditioning theory, or on Skinner's principles of operant behaviour, incorporated an intervening variable. It has not been clear whether the intervening variable undergoes a decrement during vigilance. Neurological responses, such as evoked cortical potentials, have been proposed as indicators of the intervening variables but these physical measurements are only partially interpretable. These measurements become mere input-output statements and do not allow any finer dissection of the parts and mechanisms of the vigilance process (Loeb and Alluisi, 1984).

5.1.2 Neurological models

There are two neurological models that have been used in understanding vigilance decrement. Properties ascribed to arousal systems and the neuronal property of habituation have both been invoked in explaining changes in the level of vigilance.

Arousal The neural excitation continuum view of arousal (4.2.1) has been used in explaining vigilance decrement. Thus, the decrement is due either to a gradual shift down the arousal continuum or to an habituation of the arousal mechanism (Sharpless and Jasper, 1956; Mackworth, 1970). There are a variety of observations which support the suggestion that arousal mechanisms can be intimately involved with the vigilance process. Drug effects from both depressants and stimulants produce the predicted changes in vigilance performance. However, drug effects are systemic and changes in expectancy, response criteria or other related but unobservable entities might be mediating the vigilance process. A variety of other activities such as irrelevant stimulation, mild exercise and low-level noise also alter vigilance consistent with their known effects on altering arousal (Loeb and Alluisi, 1984). Just as there are many observations supporting the arousal concept, there appear to be as many suggesting that the arousal mechanism in isolation is insufficient to explain the vigilance process (Hawkes, Meighan and Alluisi, 1964). Vigilance decrements occur even when subjects are involved in multiple tasks and arousal levels are high. Vigilance changes which are associated with sleeplessness do not fit a global concept of an arousal effect. The arousal mechanism, its strengths, weaknesses, present controversies, measurement and role in the vigilance process are considered later (5.4-5.5).

Neural habituation A variety of behavioural experiments have presented conflicting evidence for the presence of neural habituation in man and the place of neural habituation in the vigilance decrement. While it has been possible to observe a decrease in the neural response of the central nervous system sensory nuclei with repetitive stimuli, the transposition of these physiological observations across species to changes during human behavioural responses is a questionable exercise.

5.1.3 Psychophysical or information processing models

There are similarities between the neural habituation theory, learning theories and Broadbent's filter theory of attention (Broadbent, 1958). The basic property of this filter is that some information is preferentially passed onwards for further processing. In time, extra effort becomes necessary to ensure the attentional focus is maintained as a relocation of processing resources becomes likely, because of the effect of other inputs.

Another psychophysical model depends upon expectancy theory (Baker, 1959), where the difference in the expectations of subjects with respect to the temporal pattern of a stimulus and when the stimulus actually occurs is a significant factor in producing a vigilance decrement. Once again, the basic mechanisms are unknown and unobservable and interpretations of expectancy effects are argued about.

Expectancy theory has been restated in the concept of a probability-matching mechanism. Craig (1976) suggested that the decrement occurs because subjects respond at a higher rate than the signal rate, but in time reduce their response rate to match the signal rate. It is uncertain whether additional processes such as non-observing are occurring. Another possibility is that the hit (success) rate has been increased a little but the false alarm rate has been decreased considerably.

Models of signal processing which have been derived from information theory have also been used to explain some features of the vigilance response. Information theory has been applied particularly in the area of channel capacity where the vigilance decrement has been attributed to capacity being exceeded such that the facilitating effects of arousal are neutralised (Hawkes et al., 1964). While explanations related to memory load (Parasuraman, 1979) and the type of processing mode are interesting (Fisk and Schneider, 1981), they are less applicable to the monitoring situation in the cockpit.

It has been suggested that the vigilance decrement is due to a criterion shift. There has been experimental work which supports a progressive increase in conservatism of responses as a function of the duration over which responses are occurring. This has been analysed in terms of the theory of signal detection (TSD). The TSD interpretation of vigilance responses has received support from a variety of studies, but there has been methodological criticisms about the application of the theory to the vigilance situation (Loeb and Alluisi, 1984), particularly using the indices that are provided by the TSD theory.

5.1.4 Attentional systems

Much of the research in attention is directed at the process of selection, its consequences in the sensory system, or input level, as well as in the preparation of the body for change. The proposition that peripheral sensory modulation might act as an attention regulating mechanism in humans has received some support from animal experiments, although this physiological attentional filter need not be Broadbent's psychologically based filter (Hillyard and Hansen, 1986). Human electrophysiological studies using event related potentials recorded over the cerebral cortex suggest that there are attentional channels which selectively process the set of stimuli which are being attended. It is possible to delve further into conceptual schemes of the attentional

Vigilance Mechanisms

mechanism, because there are alterations in several control systems which appear to prepare the body for change. Thus, the 'orienting response' has been viewed as an attentional process. The relevance of these systems to those responsible for sustaining attention is unknown.

5.1.5 Summary

It is not possible to determine how many of the vigilance research results can be applied to the attentional process in the cockpit. A particular area of concern is the very low chance of an event which must be detected and the very long periods that vigilance must be maintained. Most vigilance research examines the vigilance decrement with respect to high response rates over short durations, rarely longer than an hour (Loeb and Alluisi, 1984). It has been suggested that it is very unlikely that a single mechanism can and will explain changes in vigilance in workplace situations, and a variety of approaches are probably able to make similar conclusions.

Older learning theories have been criticised because they degenerate into simple input-output statements rather than being testable theories. Newer theories are not beyond reproach in this regard. What is apparent is that if one, part of one, or more than one of the theories are to be used for understanding and managing the vigilance process in the cockpit, then caution is required both in the examination of data and in its interpretation. Verifying model components and quantifying predictions in the cockpit simulator may be difficult. The uniqueness of the pilot-machine relationship and the doubts surrounding the validity of the simulator environment (Roscoe, 1991) suggest that simulator observations should be used very cautiously.

5.2 Role of arousal

Arousal and its relationships with performance and vigilance have been discussed at length by many but agreement on the various constructs involving these entities still seems distant (Eysenck, 1982; Sanders, 1983; Parasuraman, 1984; Loeb and Alluisi, 1984; Hancock and Warm, 1989; Neiss, 1988; 1990; Anderson, 1990). With respect to the cockpit, arousal and its relationship to performance has been used to explain changes in crew function (O'Hare and Roscoe, 1990; Green, Muir,

James, Gradwell and Green, 1991), although not particularly in the ability of crew to remain vigilant. The relationships between arousal, performance and vigilance will be considered in some detail, starting with those that exist between arousal and performance. Some understanding of the various views and issues surrounding this relationship is required before the structures responsible for the arousal and vigilance mechanisms are considered. Arousal, and its relationship to stress, has already been discussed (4.2).

5.2.1 Arousal performance relationships

There have been two hypotheses regarding the relationship between arousal and motor performance, namely drive theory (Hull, 1943) and the inverted-U hypothesis (Yerkes and Dodson, 1908). The former is not in favour because it has not received real empirical support, it has been difficult to test and there is anecdotal evidence that the performance of some, like athletes, musicians and dancers, declines with excessive arousal rather than increases. Thus, most have favoured the inverted-U hypothesis, which asserts that there is a curvilinear relationship between arousal and performance such that performance increases with increasing arousal until it plateaus and then declines. The inverted-U hypothesis has been used in discussing cockpit arousal performance relationships (O'Hare and Roscoe, 1990).

The inverted-U relationship has been reformulated by relating the optimal range of arousal to the nature of the task (Easterbrook, 1959). For simple tasks with few critical cues, performance generally improves during emotional arousal, the range of arousal effects being large. In contrast, there is a narrowing of the range of cues used during complex tasks, performance peaking with minimal emotional arousal. Narrowing of the range of cues with complex tasks has been described (Jennings, 1986). Higher arousal levels result in performance decreases and an inverted-U relationship.

There is a broad consensus that there are levels of arousal where small increases and or decreases in arousal produce similar changes in performance. This type of trend is common to a multitude of tasks, although the slope of the relationship between arousal and performance is likely to be specific to the task, the neurobiological state and the arousal level. This view of the relationship is not at odds with observations made in neurophysiological, neuroendocrine or endocrine research studies. The controversial feature of the relationship between performance and arousal has been the decrement in performance as arousal level has increased. There are many explanations for the overarousal effect.

5.2.2 Overarousal mechanism

Explanations for the decline in performance with an increase in arousal have included a narrowing of attention from a reduction in the range of usable cues, the promotion of competing responses by the increase in arousal and a distraction effect from increased arousal producing an attentional shift to irrelevant cues (Jennings, 1986). There has been little agreement about whether overarousal exists and whether it has effects on arousal and thus on performance. This lack of agreement is easily illustrated using the effects of noise on arousal and performance.

The effect of noise has been extensively studied without clear agreement as to its effect on performance. One view is that noise-induced arousal produces specific attentional changes, mixed with other effects. These include the promoting of lapses in performance from attention diversion and an increase in high certainty responses as secondary cues are eliminated (Broadbent, 1978). An alternative view is that noise-induced arousal has consistent performance enhancing effects which are relatively non-specific but there is a coexistent masking phenomenon which produces a noise related decrement (Poulton, 1979). Jennings (1986) has pointed out that these studies are flawed because arousal is not measured and the influence of task with respect to performance is relatively uncontrolled, particularly as the duration of study lengthens. The failure to measure arousal raises the possibility that noise has altered performance, but fluctuations in arousal have had no causal role. The linking of phasic and long term physiological changes with each other and with arousal system activity is likely to be important, but these links have been ignored.

5.2.3 Inverted U-hypothesis consequences

The inverted-U relationship between performance and arousal is too simplistic in many respects. Neither arousal nor performance are specific entities, each being complex and unrelated to specific systems or measurable items. The U relationship, inverted or otherwise, must encourage thinking in terms of an optimum level, which may have no real basis. The inverted-U relationship abets other concepts which, despite their intuitive attractiveness, significantly obstruct further understanding. Two of these are the concept of global arousal and the causal or correlational nature of the hypothesis. Both these concepts will be considered as they have been used to explain cockpit events.

Global arousal It has become apparent that the unitary construct of global arousal is not useful. In most situations, indicators of arousal are poorly correlated, function independently of one another and are patterned by the current affective state (Neiss, 1988). When much of the experimental arousal data has been reexamined, it has been possible to distinguish several separate interacting neural systems which have different roles. Some systems regulate the arousal from input, while others coordinate arousal and activation (Pribram and McGuinness, 1975). Others have gone further, discriminating different arousal regulatory systems (Tucker and Williamson, 1984). It has been suggested that outside any consideration of biochemical and physiological measurements, it is possible to differentiate global arousal into arousal associated with fear, anger, excitement, sexuality and even sadness (Averill, 1969; Schwartz, Weinberger and Singer, 1981). The concept of undifferentiated arousal is not useful in general, or in the context of vigilance (Parasuraman, 1984), and should not be part of any application because it disregards psychobiological states (Neiss, 1988).

Causal-correlational issues There is controversy about the causal nature of the inverted-U hypothesis. Some have claimed that excessive arousal causes performance decrement, but others have pointed out that arousal depends upon a definition of input and output systems, it not being a naturally occurring state in its own right (Neiss, 1988). Rather, it is a variable that changes in the context of one or other of several psychobiological states (Jennings, 1986; Neiss, 1988). Thus, a cautious view of the inverted-U hypothesis suggests that the hypothesis only proposes the shape of the relationship without explaining what internal state or processes are involved. It is possible that the concept of arousal acting as a key intermediary with onward influences is incorrect. Thus, changes in arousal may be similar to heat generated by a light globe, or noise generated by an imperfect machine. They are all 'bystander' phenomena. Arousal is reconsidered later (5.4).

5.2.4 Summary

While arousal defined as a neuro-excitatory continuum is a useful concept for understanding the underlying processes, it is the relationship of arousal with other entities, such as attention, performance, task and vigilance that produces significant problems. The neuro-excitatory continuum can reasonably be ascribed to the sum of activities in different functional components of a neural network (5.4). In this context, the issues that have caused so much trouble, such as the global view of arousal and the relationship between arousal and performance, become a little more manageable. The neural network approach is not perturbed by the global issue for this is resolvable by the extent to which parts of the neural network become synchronously engaged during different neurobiological and neuropsychological states. Similarly, the causal-correlational issue can be understood in terms of neural network changes which bring about meaningful physiological alterations.

The approach which views the neuro-excitatory continuum as the sum of activity in parts of a neural network (5.4) will not satisfy all, for it assumes that overall system performance is understandable by focusing on the performance of individual parts. This is improbable as system performance is likely to be more than the sum of component performances. Nevertheless, for practical applications there is much to recommend considering arousal as an aggregate of drives in different parts of a neural network.

Overarousal remains a difficult area. Saturation effects are common in neural drives, particularly in the autonomic nervous system. Some parts of the autonomic nervous system network will limit performance as neural drive increases, producing a degree of asynchronous activation in different parts of the network. Asynchronous activation in some circumstances promotes a decline in network performance and could be the basis for an overarousal effect. Thus, monitoring management systems which measure arousal will have to be capable of assessing the import of two arousal levels for most performance levels.

5.3 Vigilance system components

There are areas of the nervous system which appear to be intimately involved in vigilance processes. Despite considerable variation in methodologies, different studies have demonstrated that specific central nervous system regions are consistently implicated. Some models of attention mechanisms (Pribram and McGuinness, 1975; Sanders, 1983; Hancock and Warm, 1989) are also of conceptual interest, despite doubts about their applicability (Parasuraman, 1984; Hancock and Warm, 1989).

The human nervous system is complex. The approach used in this book is to assume that the reader has little prior knowledge of neurophysiology or neuroanatomy. Enough detail will be given in each example such that the context in which information has been obtained is understood. Each example will illustrate some aspect of vigilance system structure and function, there being no attempt to be all inclusive. Where appropriate, methods will be briefly considered because of their relevance to vigilance measurements (7.2-7.6).

The emphasis in the following examples is on understanding the mechanisms underlying vigilance or sustained attention rather than those responsible for the onset of the attentional state. However, there are processes subserving attentional selection which may also be important. Thus, processes which stop distracting effects may decline functionally in time. Ideally, there are two requirements for assuming that a neural component is part of the vigilance system. Firstly, the activity level of the neural component must change in line with the change in the level of vigilance, and secondly, there must be some correlation with change in performance. These requirements have only been partly met in research studies.

5.3.1 Cortical systems

The varied methodologies used in advancing our knowledge of human attentional systems have included illness presentations, cortical neurophysiological studies, cerebral metabolic and blood flow studies, and psychophysiological studies.

Clinical The clinical world has always been a rich resource with respect to understanding man. Clinical neurology, with its descriptive if not curative skills, can be very useful in delineating those parts of the nervous system important in vigilance. The practice of using 'nature's experiments' to enhance knowledge has been justified by its contribution for all rather than for the individual.

Vigilance Mechanisms

Spatial neglect is the unawareness, often coupled with a failure to respond, to stimuli in specific locations in space. Neglect following disease induced brain damage usually occurs in a sector of the space around the subject which is on the opposite side to the damaged part of the brain. Damage to certain parts of the brain bilaterally produces neglect in more than the horizontal dimension. Thus, bilateral posterior parietal lobe damage has been associated with attentional defects in radial space, suggesting that neglect can occur in multiple spatial dimensions and that there is a three-dimensional attentional system in humans (Mennemeier, Wertman and Heilman, 1992). Attentional defects are complex. Some have suggested that elements in the area of neglect are not ignored. Rather, the space in which they were in becomes distorted, such that the elements become compressed non-linearly in the remaining attended areas of space (Halligan and Marshall, 1991). Damage to other areas of the brain can produce auditory attentional neglect. The consequences of a variety of neurosurgical procedures used in epilepsy and neuropsychiatric therapy provide similar information about the attentional roles of different cortical areas.

In summary, it is likely that there are discrete areas of the brain which either house essential circuits or through which essential interconnecting systems run, all having a functional role in sustained attention. It is likely that there are different circuits for different environmental stimuli and that the attentional processes may be coordinated across some modalities but not across others (Inhoff Rafal and Posner, 1992).

Neurophysiological It is possible to record electrical changes (7.2) in many regions of the human brain by recording from the scalp. There are several event related brain potentials (ERP), ranging from very short latency waves localised to the brainstem (Jewett and Williston, 1971) to much longer latency waves (Walter, Cooper, Aldridge, McCallum and Winter, 1964). Debate centres on which neural system is responsible for these potentials, it being likely that they are produced by a number of systems working together. Changes in ERP have been observed in the course of vigilance tasks of different types (Haider, Spong and Lindsley, 1964), vigilance decrement being associated with a reduction in electrocortical arousal (Davies and Parasuraman, 1982). Because some components of ERP increase when attention is directed to specific inputs, it has been possible to distinguish between selective and sustained attention at a brain potential level. Cognitive components in

ERP can be localised as dipoles in different cortical areas. ERP studies suggest that there are complex interactions between the posterior parietal and prefrontal cortex during attention (Desmedt and Tomberg, 1989). It is now possible to assign a spatial orientation and a vector magnitude to dipoles, dipoles being distinguishable upon characteristics of the attentional state.

Cerebral blood flow Positron emission tomography (PET) allows real time detection of blood flow changes in different parts of the brain. PET studies show that visual imagery, word reading and shifting visual attention from one location to another are not performed by any one brain area. Selective attention uses separate systems from those that passively collect information about a stimulus (Posner, Petersen, Fox and Raichle, 1988). Some studies have shown that for attentional processing there is a mid-line attentional system in a region of the anterior cingulate cortex (Posner et al., 1988; Pardo, Pardo, Janer and Raichle, 1990). Simple visual and somato-sensory vigilance tasks are associated with increases in blood flow in the prefrontal and superior parietal cortex in the right hemisphere, regardless of the modality or laterality of the input (Pardo, Fox and Raichle, 1991). It is this area of the cortex that if damaged is associated with inattentiveness to personal space (5.3.1). There is an overlap in these observations with those from electrocortical recording and from vigilance research on cerebral laterality (Parasuraman, 1984).

Summary There are several attentional systems that manage incoming information in different sensory systems. Relatively localised right hemisphere areas are used by some sensory inputs while a mid-line attentional system is used by others, both hemispheres being required for the full performance of the whole attentional mechanism. Although these conclusions apply largely to cortical areas it is likely that much deeper central processing structures, such as the deeper cortical regions and the basal ganglia also have roles in attentional systems. These cortical systems satisfy the conditions necessary for inclusion in the vigilance system in that there is some evidence that neural activity increases with vigilance level because neuronal cell masses become metabolically more active. There are many examples where absence of neural activity in these areas is associated with reduced vigilance performance.

5.3.2 Reticular activating system

The human cortex is relatively large and focal damage, produced by a disease process or a neurosurgical procedure, can provide information about specific nervous system functions. Similarly, cerebral blood flow studies, other imaging approaches and neurophysiological techniques have allowed us to make functional observations on cortical zones because cortical and subcortical areas have relatively large volumes. These techniques become much less capable in the lower brain. The lower brain has considerable processing capacity in its own right and contains a multitude of neural pathways travelling to and from the massive cortex above it. All these functional networks are relatively compressed and thus unsuitable for the techniques that have been useful in understanding cortical function.

The brainstem has a multitude of functions and many of these use collections of nerve cells and nerve fibres which are located in a region called the reticular formation. This is particularly inaccessible to all the techniques described above, direct measurement of reticular formation function not being possible in man. While experimental work has provided some insight into the importance of the reticular formation in the regulation of arousal, its role in sustained attention is unclear. Nevertheless, there is probably an extended neural network important in the control of adaptive behaviour, which involves the ascending reticular activating system and its interconnections with the frontal cortex (Vanderwolf and Robinson, 1981; Parasuraman, 1984; Vanderwolf, 1992). This system may have involvement in controlling the level of sustained attention as well. The reticular formation is part of the brainstem and midbrain, having a fundamental role in a large array of essential body functions. Damage to parts of the reticular formation is associated with disorder of basic body functions and poor performance on a continuous performance test, suggesting that the reticular formation has an attentional role (Parasuraman, 1984). In contrast with the cortex (5.3.1), the reticular formation does not satisfy the preconditions required for inclusion in the vigilance system (5.3). This may be because it is so difficult to measure reticular neuronal activity directly.

5.3.3 Autonomic nervous system

It has not been possible to measure the direct contribution of the reticular formation to the vigilance process, indirect methods being required to asses this complex controller and system regulator. In the reticular formation lies neural circuitry important for the control of blood pressure, respiration, digestion, excretion, reproduction, thermoregulation, the sleep cycle and arousal. In vigilance research studies it has become common practice to measure one of the related variables such as heart rate, even though heart rate has a primary role in blood pressure regulation. This indirect approach assumes that the measurement of the related variable gives an indication of the activity level of other reticular system networks, including those subserving vigilance processes. This assumption reflects the belief that vigilance requires alterations in the related variables for its full expression (5.3.4).

It is debatable how useful these related variables are. For example, much of the variation in heart rate has nothing to do with changes in vigilance (Jorna, 1991), being dominated by respiratory regulatory inputs, thermoregulatory inputs and by the needs of the blood pressure control system. Under very controlled conditions, some changes in heart rate may partly reflect the levels of activity in neural systems due to workload and fatigue as well as system drives subserving effort and attentional selection (7.7). Undoubtedly, it is very difficult to measure reticular system function directly. The desire to measure a reticular system output in vigilance studies has been great, and with a few exceptions there has been little questioning of the validity and utility of using autonomic nervous system measurement for this purpose.

The ease of measurement of autonomic nervous system outputs is great as the reticular system connects to visceral organs like the heart, sweat glands and blood vessels by specific nerve fibres in the peripheral nervous system. The functions of these organs can be assessed with relatively robust, reliable methods. The peripheral nerves are very distant from reticular and cortical areas important in sustained attention, and the nature of the interconnecting pathways is such that attentional system signals are open to modulation at a number of sights. Measures that indicate that part of the autonomic nervous system is engaged in some way during sustained attention are numerous, but the autonomic nervous system's role in vigilance processes is far from clear.

The simplest view, consistent with the preconditions required for inclusion of neural systems in the vigilance process (5.3), is that the autonomic nervous system is only a conduit and reveals reticular system activation during vigilance in a very imprecise way.

5.3.4 Vigilance associated outputs

The changes in a reticular system output like heart rate during exercise are relatively well understood and non-controversial. Explanations for the change in heart rate during attentional selection have been considered extensively but without clear agreement (Jennings, 1986). In contrast, the changes in reticular system function during sustained attention have remained relatively unexplained and rarely considered. These are considered later (7.3-7.6).

5.4 Arousal system components

It is an assumption that there is an arousal system. This implies that there is a network of neurones which are principally involved with the regulation of the level of activation of all other neurones. It is possible that the arousal system is not a set of specific neurones but is a property of the patterns of interconnections in a number of neural networks.

Arousal systems are closely intertwined with vigilance systems (Pribram and McGuinness, 1975) and the neural network responsible for arousal, even in its differentiated form (Sanders, 1983), is likely to be involved in part in subserving sustained attention. Some have used arousal and attention interchangeably as if they are the same process. It is likely that arousal and attention are discrete entities, though they are intimately intertwined under many circumstances.

In this section the evolution of our understanding of the arousal system is briefly considered. Present concepts, including circadian rhythm mechanisms and the linking between other systems and arousal mechanisms, will be discussed.

5.4.1 Historical considerations

In the past, the terms arousal and stress were used interchangeably by researchers, current views (4.2.1) not being considered. The neural pathways underlying the arousal system were first studied by stress researchers, who adopted an animal based paradigm in conceptualising stress mechanisms. The stress process has been viewed in terms of 'fight' or 'flight' responses (Matteson and Ivancevich, 1987; Mayes and Ganster, 1988). Fight responses translate easily into political behaviour (voice, fight or attack response) as do flight responses (exit, flight or withdrawal response), both being measurable entities (Mayes and Ganster, 1988). The desire to view stress in terms of flight or fight responses still pervades much current research.

The hormones involved in fight or flight stress, namely adrenalin and cortisol, have been measured with variable results in a variety of circumstances. An arousal increasing procedure, the Stroop Colour Word test, is associated with increased heart rate, blood pressure, adrenalin, noradrenalin and respiration rate as well as changes in galvanic skin resistance and vascular resistance (Hjemdahl, Freyschuss, Juhlin-Dannfelt and Linde, 1984; Tulen, Moleman, Van Steenis and Boomsma, 1989). These changes have all been interpreted in the light of the classical 'defence' reaction pattern.

While there are central nervous system regions that orchestrate the lower areas of the autonomic nervous system into defence reactions in experimental studies, the wholesale transfer of relatively simple autonomic motor programs, such as fight or flight, from the animal kingdom to everyday human behaviour is unwise. While there may be rare instances where these reactions occur, it is too simplistic to see most environmental interactions in terms of these responses, activated to variable degrees.

5.4.2 Hypothalamic systems

Some regions of the brain, such as the hypothalamus, have specific roles in controlling temperature, blood pressure, the sleep cycle, hunger, thirst and many other entities. The hypothalamus is specialised for mixing information from the viscera, areas in the brainstem and higher centres, enabling it to regulate a plethora of neural, neuroendocrine and endocrine functions. Many of the hypothalamic control systems are ingeniously modified for specific roles in different species. Many autonomic pathways which produce unwanted physiological changes in 'stressed' individuals are partly controlled by hypothalamic systems. The hypothalamic systems which regulate temperature and the sleep cycle are important in understanding arousal in humans.

Temperature There are regions of the hypothalamus which have specific and prominent roles in temperature regulation. Shivering, sweating and the regulation of skin blood vessel diameter are the motor systems which the hypothalamus controls and integrates while combating environmental fluctuations and producing endogenous temperature changes. Body temperature is not constant, having a characteristic circadian fluctuation, being highest mid evening and lowest in the early morning. There are a variety of temperature regulating abnormalities in diseases which affect the hypothalamus.

Sleep cycle The hypothalamus is important in sleep cycle regulation and specific regions, some close to the areas involved in temperature regulation, have powerful effects on the propensity for sleep. The propensity for sleep, which varies widely between individuals, can be measured. Some hypothalamic disorders are characterised by persistent somnolence or paroxysmal attacks of sleep.

Thermoregulation and sleep cycle regulation are functionally intertwined. Normally the sleep-wakefulness cycle fluctuates as body temperature fluctuates (Glotzbach and Heller, 1989; Czeisler and Jewett, 1990). Desynchronisation of these coupled systems has prominent effects on sleep propensity and character (Glotzbach and Heller, 1989). Naturally occurring examples include the changes in temperature regulation associated with hibernation and the phase shift in the temperature cycle which has been observed in sleep-onset insomniacs (Morris, Lack and Dawson, 1990).

Bilateral damage of the major thermo-sensitive region of the mammalian brain in the anterior part of the hypothalamus results in severe and persistent reductions in sleep. The neurones in the anterior hypothalamus which are thermo-sensitive participate in the continuous regulation of both temperature and sleep propensity (Spyer, 1988). Some have suggested that when one of the 'waking' regulating systems in the hypothalamus is damaged, others readily compensate for the loss (Denoyer, Sallanon, Buda, Kitahama and Jouvet, 1991).

Arousal There is no part of the hypothalamus that is exclusively committed to the regulation of arousal, although some parts contribute significantly. However, there are parts of the hypothalamus which are involved in the regulation of the adrenal medulla (adrenalin and noradrenalin) and the adrenal cortex (cortisol) as well as the regulation of the cardiovascular, respiratory and thermoregulatory systems. Many of these variables change as the level of arousal changes. Under extreme conditions, or in the controlled environment of the laboratory, these autonomic regulatory systems can produce stereotyped responses (5.4.1).

Ablation experiments, where nerve cells and nerve fibres are damaged, have been misleading as to the role of the hypothalamus in arousal regulation. With the advent of experimental tools that allow nerve cells to be excited or inhibited without affecting nerve fibres, it has become obvious that many hypothalamic systems are part of much wider neural networks. Arousal regulation is unlikely to reside in the hypothalamus, arousal level depending upon the engagement of different networks in different ways depending upon the neurobiological state.

There is some crude evidence for the engagement of different neural networks. Pupillary diameter is controlled by both the parasympathetic and sympathetic divisions of the autonomic nervous system and changes in diameter, which are time locked to a task, give a measure of the momentary effort required. The task evoked change in pupillary diameter declines over time (Beatty, 1982), but the basal pupillary diameter, which reflects a more general arousal effect (Yoss, Moyer and Hollenhorst, 1970) does not. Thus, studies of sustained attention during a high event rate task reveal multiple underlying processes.

5.4.3 Cortical systems

Cortical systems modulate a variety of autonomic nervous system components. Those pathways that become active when there is a change in the level of arousal have not been clearly delineated. Cortical areas involved in blood pressure regulation can give some idea of the complexity of an arousal controlling cortico-visceral system. Cortical areas which affect blood pressure regulation consist of the infralimbic area of the medial prefrontal cortex, the anterior cingulate cortex and the insular cortex. These have pathways to the hypothalamus, brainstem and spinal cord. The arousal motor system may use a similar network.

The structures involved in cortical activation are unknown. The ascending system has multiple paths and a great deal of in-built redundancy. These systems spread from the brainstem to the cortex, changes in one system having immediate repercussions on others (Denoyer et al., 1991).

5.4.4 System linkages

It is possible that arousal control resides in the way neurones are interconnected and is a property of a network of neurones rather than being controlled by a specific set of nerve cells. Thus, the arousal system resides in the interactions of other networks, these interactions being dominated by the characteristics of the networks and the temporal patterning of their interaction.

Network type There are complex interactions between many of the networks in the autonomic nervous system. Networks like the thermoregulatory and sleep cycle systems have very predictable interactions (5.4.2).

Temporal pattern Some interactions are quite plastic in time, the linkage between systems having significant historical features. Activation of the adrenomedullary system, with release of adrenalin and noradrenalin from the adrenal medulla, has consequences in other autonomic nervous system pathways. Thus, an infusion of adrenalin, which mimics levels in man which occur during stressful situations, is associated with a continuous discharge in sympathetic nerve fibres long after the infusion has stopped (Persson, Andersson, Hjemdahl, Wysocki, Agerwall and Wallin, 1989). The duration of sustained nerve firing increases with successive bursts of adrenalin. Thus, the responsiveness of some systems will reflect in a unique way the recent (hours, days or longer) fluctuations of activity in other neural and neuroendocrine systems. Other recent studies in animals support interactions between the 'arousal' system and other autonomic systems (DeBoer, Koopmans, Slangen and Van der Gugten, 1989) as do studies outside laboratories (Balick and Herd, 1986).

5.4.4 Summary

It is uncertain if an arousal system exists. Forms that arousal may take include:

1 Arousal is only system noise. Arousal is a byproduct of the activity of other neural control systems and has no causal role (5.2.3).

2 Arousal is a property of a neural network. Here, arousal is a plastic entity, partly characterised by whichever system has

changed its level of activation. The plasticity also applies to temporal properties, amplification or accommodation being prominent features. This form has the output of the relevant system shaped in a beneficial way by the presence of the arousal effect.

3   Arousal is the output of a specific neural network which has a labyrinth of interconnections with most other systems. It has not been possible to identify such a network. Networks involved with sleep cycle regulation are possible candidates, although at present these do not satisfy the criteria for involvement in vigilance or arousal regulation systems (5.3).

5.5 Arousal vigilance interactions

Some have used arousal and sustained attention as if they are the same process. It is likely that arousal and attention are discrete entities, although they are intimately intertwined under many circumstances. PET studies have show that the central attentional system, activated by an arousal increasing procedure (Pardo et al., 1990), is not engaged by a task demanding sustained attention (Pardo et al., 1991).

Although different psychobiological states are characterised by different patterns of nervous system involvement, it is still instructive to look at the association of global arousal (5.2.3) with vigilance as the effect of specific affective states cannot be considered yet. There is a problem with the total rejection of the concept of global arousal because of its prior broad acceptance.

5.5.1 Global arousal

Detection efficiency in a vigilance task is reduced by moderate heat, which lowers arousal, while stimulants like amphetamines (Mackworth, 1965) and low frequency vibration increase arousal and also improve vigilance. Individuals with different levels of chronic arousal, because of personality type or mental illness, perform vigilance tasks in a manner which produces some matching between the changes in vigilance and the level of arousal. Similarly, circadian rhythm fluctuations in arousal are associated with appropriate changes in the performance of vigilance tasks (Parasuraman, 1984). However, the relationship between global arousal and vigilance is less clear cut in other circumstances.

Similar electrocortical arousal decreases occur when there is a vigilance decrement, when there is no change in vigilance, when there is a vigilance decrement linked to a high but not low target probability, when there is vigilance decrement associated with criterion change and when the subject relaxes and does not carry out a vigilance task. These studies illustrate both the causal-correlational dilemma (5.2.3) and the weakness of the global arousal construct (5.2.3). In addition, the concept of global arousal is inadequate when the latencies of positive and negative responses are measured in a vigilance task. The former lengthens as the task progresses but the latter does not (Parasuraman and Davies, 1976), this differential trend being difficult to reconcile as the speed of detection of responses is a monotonic function of arousal level.

5.6 Summary

There are a number of dilemmas in the prevailing views of arousal, vigilance and the vigilance decrement which make it unwise to use any one theory in a practical vigilance problem (Loeb and Alluisi, 1984). Almost all the vigilance studies have concentrated on vigilance tasks where the periods of vigilance have rarely been more than one hour and where signals have been presented relatively frequently. When the task becomes longer and the chance of a signal becomes relatively infrequent, much of the vigilance research information is difficult to apply. It is possible that in such situations, typified by cockpit monitoring, there is no vigilance decrement, although this is unlikely. One fact is certain, which is that in applied problems like cockpit monitoring, vigilance performance will have to be studied in situ to ensure that theoretical constructs found in laboratories and simulators are useful.

Global arousal and other constructs are suspect in a specific application (Neiss, 1988), because the psychobiological state associated with flying a modern commercial aircraft will be missed (Hancock and Warm, 1989; Hancock, 1991). It is possible that each monitoring task, whether it be in the cockpit, on the bridge of a supertanker, or in a nuclear power station, is quite specific and that great caution must be exercised in applying any general methodology. While arousal is fashionable in understanding attention, arousal regulation may well be a myth because arousal dynamics are situation specific and sensitive to recent and prior events.

Those areas of the nervous system involved in vigilance are likely to be widespread, are likely to have great redundancy which reflects fundamental survival issues, and are likely to use other parts of the nervous system for expression. The fact that there has been little need for humans to signal to the outside world how vigilant they are means that measuring vigilance will always be by indirect means.

There are areas in the parietal and prefrontal cortex that are key parts of the neural network subserving the vigilance process. The cortical vigilance neural networks are interconnected with the reticular activating system through which body structures can be engaged as required for specific vigilance tasks. The essential and similar nature of vigilance and arousal supports close linking between their respective neural networks. If arousal systems do not exist as physical structures but as properties of other neural networks, it is likely that arousal level change is a prominent feature of vigilance system activation.

6 Automation, peripheralisation, vigilance and stress

The processes underlying peripheralisation can now be detailed. It is proposed that an important element linking peripheralisation and the errors due to human factors is degradation in the quality and quantity of vigilance. The vulnerability of vigilance processes to cognitive stress is discussed.

6.1 Peripheralisation-vigilance cycle

The interaction between automation, peripheralisation and human errors has already been described (Figure 2.3). In many of the human factors accidents, including those in faulty (3.4.1) and faultless aircraft (3.4.2), unsatisfactory vigilance levels were present. Inadequate vigilance appears to have been an important element in these types of human error, occurring as a direct consequence of peripheralisation, as well as a consequence of other peripheralisation effects such as complacency, inadequate communication and reduced situational awareness. The properties of the peripheralisation-vigilance cycle (Figure 6.1) are similar to those described previously (2.6.1).

The effects of complacency on vigilance processes have not been measured, measurement of both entities being unsatisfactory at present. Measurement problems also apply to the relationships between vigilance and situational awareness.

6.2 Stress and vigilance

The previous consideration of stress in the cockpit concluded that the present position of pilots is an invidious one and is not being adequately managed (4.5). Pilots are being asked to act as monitors in situations where the monitoring role is increasingly task- rather than human-centred. In addition, their training and much of their role expectation has

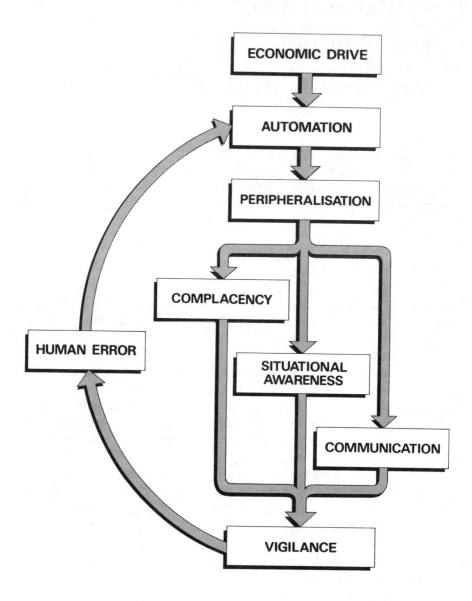

Figure 6.1 The interaction between human error, automation, peripheralisation and vigilance. Vigilance can be affected at multiple sites between peripheralisation and error

been focused on flying and not on monitoring. Pilots are being expected to act as a fail-safe device, to take over from automated systems in circumstances where the task may already have reached airframe and control system limits. Thus, there are new sources of stress related to role change and the act of monitoring that are likely to become increasingly important while the current philosophy underlying cockpit automation remains unchallenged.

6.2.1 Role ambiguity as a stressor

Strain is likely for pilots because of role ambiguity (4.3.2). Strain is not induced by stressors in the external environment, like noise or thermal stress, but by the cognitive stress of role ambiguity (4.3.2). Another form of cognitive stress is the vigilance task itself (Hancock and Warm, 1989; Hancock, 1991). It is likely that these cognitive stressors together have deleterious effects on the ability to sustain attention.

6.2.2 Vigilance as a stressor

The vigilance task itself produces stress. A number of studies have suggested that significant stress accrues when a highly demanding discrimination task is coupled with an impoverished display of information (Hancock and Warm, 1989). Adrenalin and noradrenalin responses support the stressful effect of sustained attention (Frankenhaeuser, Nordheden, Myrsten and Post, 1971) and subjective ratings before and after vigilance tasks show higher levels of fatigue and drowsiness post task (Hovanitz, Chin and Warm, 1989; Hancock and Warm, 1989). It appears as if the process of vigilance has the potential to damage itself. It has been suggested that the stress that results from a vigilance task reflects attentional demands coupled with the lack of control over the occurrence of events (Thackray, 1981). Hancock (1991) has described humans as being 'superbly disqualified for the system monitor role', being chronically overstressed when forced into it. The vigilance task is very capable of producing stress, the resulting strain being directly detrimental to the vigilance process. Thus, apart from a natural instability, which appears inherent to the vigilance process (5.6), there are other factors related to the act of sustaining attention which make monitoring a very vulnerable activity.

Cockpit Monitoring and Alerting Systems

6.3 Peripheralisation-stress cycle

It has been argued that peripheralisation produces cognitive stress by promoting role ambiguity (4.3.2), while vigilance tasks themselves are sources of cognitive stress (6.2.2). The two cognitive stressors enhance the detrimental effects of peripheralisation on monitoring (Figure 6.2).

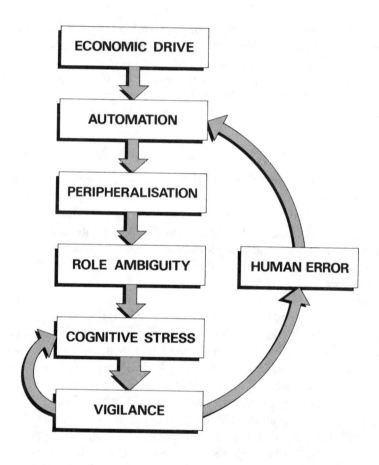

Figure 6.2 The peripheralisation-stress cycle

6.4 Multiple positive feedback loops

The monitoring role in the cockpit is beset by factors which promise to hinder humans in making their unique contributions of pattern recognition and creative problem solving. These factors interact in positive feedback loops or cycles. It is unknown how fast the peripheralisation-stress cycle or the vigilance-stress cycle are rotating, nor is it known how they are coupled.

6.5 Solutions

Any solution to the monitoring problem must accept that much of the drive for automation is economically driven. Commercial aviation organisations operate with small margins and much of their profitability comes from continually reducing seat-kilometre costs. These have dropped dramatically because of new technology, but future cost decreases will only come about because of further automation. Given that automation is to be welcomed, and the trend to peripheralisation inevitable, the solution to ineffective vigilance must involve optimising the human contribution to monitoring, while removing the factors which are damaging to it.

Ideally all damaging cyclical effects should be removed. The only way to do this and stop error production is by removing the factors which are degrading the ability to sustain attention. Neutralising stress effects, altering monitoring ability and altering the nature of the monitoring role would seem to be vital. Human change is not a possibility and human selection, while possibly useful, is only tinkering at the edges of the problem. CRM programs, useful as they are, do not promote better monitoring and their stress management programs do not target the cognitive stresses produced by vigilance and role ambiguity. It is proposed that alteration of human monitoring ability requires alerting system technology and altering the monitoring role requires changing the management of the human monitoring resource in the cockpit.

Part B
Monitoring, Measurement and Alerting Systems

7 Vigilance measurement

Vigilance has been measured in a number of ways. Most have proved to be unsatisfactory. Resolution of the vigilance measurement problem requires definition of those parts of the human nervous system which are involved in sustaining attention (5.3). Attentional selection should be distinguished from sustained attention as the monitoring role in the automated cockpit predominantly involves the latter. The focus on sustained attention must also allow for the complex linkage between vigilance and arousal (5.3-5.5).

In this section vigilance measurement processes will be described after consideration of some general issues related to measurement in the attentional-arousal areas of the nervous system. The interrelationship between vigilance, workload and effort will also be discussed.

7.1 Measurement issues

Changes in the neural networks important in vigilance and arousal are frequently affected by other neural drives. These produce noise when vigilance is measured. Other factors affecting vigilance measurement include changes in the state of neural networks and the appropriateness of the accepted measures of neural activity.

7.1.1 Sources of measurement noise

Noise is signal that is not relevant or interferes with the measurement of a desired signal. There are several types of signal noise. There is biological noise and measurement system noise, the latter representing spurious information generated at the human-machine interface or noise within the measuring device itself. This type of noise becomes more important as the signal of interest decreases in amplitude. There are a variety of techniques which can minimise the effect of this type of noise.

Biological noise can be divided into 'random' noise and 'systematic' noise. Systematic noise is noise that is related to the variable under investigation but is not produced by the relevant neural networks. Thus, event related electrical potentials (ERP) recorded from the scalp have been used to understand human attentional systems (5.3.1) but the ocular potential, recorded at the scalp, is a potent source of systematic noise. The ocular potentials are driven by the tasks in attentional studies and systematically interfere with adjacent cortical recordings. Signal subtraction techniques can reduce this type of noise. Another source of systematic noise is produced by structures adjacent to those of interest which are also perturbed to some degree by the environmental stimuli under investigation. The problem here is the limitation imposed by the focusing ability of the recording device, signal subtraction not being possible. Electro-cortical recordings, even with all their modern refinements, are still only recording small currents within a volume conductor and will always be prey to recording from adjacent but unintended targets.

Random noise is a problem in vigilance measurement because the neural output of interest is influenced by more than one input. Measurement systems which are unable to use repetitive stimulation and averaging techniques require knowledge of all inputs into the neural network of interest if the effects of this type of noise are to be minimised. A well known example is respiratory breathing rate which reflects the fundamental need for oxygen intake and carbon dioxide expulsion. However, respiration subserves the needs for speech as well. The twin drives of respiration and speech produce a respiratory breathing rate which cannot easily be correlated with either drive. Random noise is a problem in vigilance studies.

7.1.2 State change

Circadian and other rhythms are a feature of the neural networks which interact with and are part of vigilance systems (5.3-5.5). The functional state of these rhythmic networks fluctuates significantly so that measurements made in related systems can only be interpreted in the context of the conditions pertaining to the time of observation. While state setting factors can be partly controlled in vigilance research studies, it is unlikely that they can or should be controlled in the workplace environment. Some workplaces like cockpits have partly controlled environments and exert significant restrictions on personnel. Thus, variables such as temperature and noise are kept within limits while physical activity, eating and excretion are relatively programmed. However, in long distance flying the probability of significant changes in neural network state is relatively high.

7.1.3 Output measurement

Nerve cells have excitable membranes which can change their electrical state and produce impulses (action potentials). Information is encoded in the number of impulses in a unit of time and in the temporal pattern of impulses. It is quite possible that neuronal network activity is not described completely by the mean discharge rate or by the pattern of discharge, the degree of activation of the network being also reflected by the extent of the spread of activation. It is not clear that vigilance measurement takes into account other measurement variables.

7.2 Electroencephalography

Under some conditions, the electrical currents recorded from the surface of the brain correlate with activity in single neurones and groups of neurones. The currents recorded from the scalp are used to produce the electroencephalogram (EEG), which gives a representation of neural activity in some brain regions. However, it is not clear that the EEG provides measures that are sufficiently focused. Experimental work suggests that the EEG reflects activity in a complex set of endogenous circuits, with contributions from the thalamus and diffusely projecting pathways which use different neurotransmitters (Vanderwolf, 1992). Electroencephalographic measurement of vigilance level may be relatively non-specific and dominated by systematic sources of noise.

7.2.1 Arousal and vigilance studies

Many have assumed that arousal and vigilance are accurately indicated by electroencephalographic patterns. The traditional approach measures the presence of specific frequencies in the EEG. The amplitude of electrical activity in the 3-5Hz band has been taken as an indication of distinctly reduced alertness, while the amplitude of activity in the 8-

13Hz band is characteristic of alertness. Other in between states are recognised. Many studies have employed EEG measures as the benchmark of the state of arousal. Thus, many drugs which have a sedative action and many which require evidence that they have no sedative action are being assessed using EEG rhythm analysis at various drug doses.

Experimental studies suggest that the presence of low frequency activity in the EEG may not indicate a state of reduced arousal but is more indicative of a dementia-like state (Vanderwolf, 1992). Humans who exhibit slow wave patterns during the waking state have behavioural disorders, severe memory problems or automatic activity patterns, their changes in vigilance being a secondary phenomenon. Slow wave patterns during sleep are associated with sleep-walking and night terrors. Slow wave EEG rhythms occur in most people during normal sleep, the sleeping state masking dementia-like behaviour.

Event related potential studies have been used for understanding attentional selection under laboratory conditions (5.3.1). It is likely that EEG measures provide some indication of the attentional state, less than claimed by the devotees, but possibly useful in the context of a specific work place.

7.2.2 Field studies

Results from laboratory based studies suggest that EEG measures of the attentional state may suffer from being too non-specific. Field studies using EEG measurements in simulation systems have similar problems. There have been disappointing results from biofeedback control of EEG rhythms, a possible factor being non specific measurement effects (Parasuraman, 1984). The vigilance issues in car and road design are being tackled using instrumented drivers carrying out long distance driving tasks. Changes in EEG frequency content have been used to determine the effect of temperature, noise and vibration on the vigilance level of car drivers (Pettit and Tarrière, 1991). Noise has an effect on vigilance but only after several hours driving, vigilance decreasing throughout the observation period (Fakhar, Vallet, Olivier and Baez, 1992).

Recording requires electrodes being attached to skin. Long term stability is a problem, even for those with extensive recording experience. Thus, in the clinical investigation of obscure transient neurological syndromes, long term (8hrs) EEG measurement requires a considerable commitment to technique, patients being cautiously ambulant wearing the equivalent of an outsized turban. There are more aesthetic military systems but the vulnerability and fragility of the skin-electrode interface is still present. Recent attempts to use EEG measurements as correlates of decision making strategies in simulated combat required highly specialised equipment and post recording 'cleansing' of artefacts. Even then, a third of recordings were technically faulty (Poe, Suyenobu, Bolstad, Endsley and Sterman, 1991), suggesting that there are significant practical problems underlying the routine use of EEG recordings.

7.2.3 Cockpit studies

It has been claimed that advanced methods of EEG analysis may be useful in tracking cognitive functions and perceptual strategies in military cockpits (Sterman and Olff, 1991). These studies used the amount of 8-12Hz activity in certain cortical areas as a measure of the level of arousal and as an index of task demand or workload. Reasons for doubting the wisdom of this type of approach (7.1.1, 7.7) include the measurement of different neuro-behavioural entities by the same technique, stability, the validity of the data analysis techniques and the interference with the task by the measurement systems. However, the combination of multiple measures of vigilance, including the amount of power in the lower frequencies of the EEG, with a conservative interpretation of data may be a useful tool for obtaining information about vigilance in long haul flying (Fouillot, Coblentz, Cabon, Mollard and Speyer, 1991; Cabon, Mollard, Coblentz, Fouillot, Stouff and Molinier, 1991). It is the combination of EEG measures with other types of measurement which helps offset the deficiencies of EEG monitoring.

In a study designed to develop crew rotation schedules for the Airbus A340, four variables, the EEG, the electrooculogram (EOG), heart rate and wrist movement, were measured in line pilots flying commercial sectors while observers documented aircrew task and environment (Cabon et al., 1991). Vigilance, as assessed from the EEG and eye blink frequency, declined during periods characterised by a low workload and monotony. Significant vigilance decrement started within 30 minutes of take-off and was associated with a decrease in communication. This type of study is important, because the measurements, imperfect though they may be, give an indication of vigilance level in the cockpit environment itself. While the presence of observers is likely to alter cockpit dynamics, detracting from this approach, further technological and methodological improvements should provide important and useful information. At present these approaches do not lend themselves to routine cockpit operations.

7.3 Heart rate

Heart rate is dependent upon many factors. The parasympathetic and the sympathetic nerves provide the direct neural control, the former being able to alter heart rate on a beat by beat basis. These two parts of the autonomic nervous system have their controlling nerve cells in various parts of the brainstem and in ganglia. They are affected by a myriad of central nervous system drives and reflex systems, some involving other viscera. There are other indirect neural controls via adrenalin (7.5) and other circulating substances.

Heart rate has frequently been used as a physiological measure of autonomic nervous system function, both in the research laboratory and in the workplace. Heart rate can be measured as the mean heart rate or as the heart rate variability (5.3.3). Mean heart rate reflects the summed effects of a myriad of competing influences, comprising neural, hormonal and physical factors, while heart rate variability reflects more transient influences (5.3.3). Vigilance systems are only one of many systems controlling mean heart rate (5.3.3). Random and systematic noise are major factors in reducing the value of mean heart rate measurement.

Heart rate variability, which is beat to beat variation, reflects a narrower set of influences, where the sympathetic nervous system, adrenalin and noradrenalin are relatively less important, the parasympathetic vagus nerve having a dominant effect. Heart rate variability has been measured in many ways. Spectral analysis of the frequency content contained in the fluctuation in heart rate is a powerful technique. It allows the relative amount (power) of heart rate variation at each frequency to be determined. The power in different frequency ranges can reflect effects from the thermoregulatory, blood pressure and respiratory control systems (Jorna, 1991).

7.3.1 Arousal and vigilance studies

Heart rate has been measured in arousal studies (Hjemdahl et al., 1984; Tulen et al., 1989), increased arousal being associated with clear increases in mean heart rate. Heart rate remained stable or increased during a vigilance task (Beh, 1990), but was uncorrelated with performance (Parasuraman, 1984). In high achievers, heart rate increased with vigilance duration, performance remaining higher but not being tightly correlated (Beh, 1990). Heart rate variability declined during a vigilance task. There has been some similarity in the heart rate changes during arousal and during the effort to sustain attention. Heart rate increases as vigilance proceeds, either due to sympathetic activation or withdrawal of vagal parasympathetic drive or both. The decline in heart rate variability during a vigilance task has been consistent with withdrawal of vagal tone (Grossman, Stemmler and Meinhardt, 1990).

7.3.2 Field studies

Heart rate has been used in studies (8.3.1) of the attentional state of train drivers operating the paced secondary motor task (SIFA) taken from locomotives in the German Federal Railway (Peter, Cassel, Ehrig, Faust, Fuchs, Langanke, Meinzer and Pfaff, 1990a). Heart rate increased gradually as alertness, measured by EEG criteria, declined. Heart rate variability also declined. Heart rate changes have been attributed to the physiological effort required to achieve alertness sufficient for performance under monotonous conditions. Heart rate is affected by isometric exercise, a factor in field studies.

There have been significant differences between laboratory and field studies of heart rate changes during vigilance tasks (Beh, 1990; Peter et al., 1990a). In instrumented car drivers, heart rate declined as the duration of driving increased, heart rate variability showing little change (Fakhar et al., 1992), but heart rate variability has increased as driver fatigue has increased (Mackie and Wylie, 1991). There have also been significant discrepancies in the heart rate changes observed in laboratory and field studies carried out by the same observers. Heart rate changes in field studies can be useful, if the studies are well controlled, comparisons are restricted to different states within the one study and the analytic methods are appropriate.

7.3.3 Cockpit studies

Cockpit heart rate monitoring has been used for research purposes (Graeber, 1988; Thomas, 1989). Heart rate was low during cruise, with significant increases during take-off, descent and the final approach (Kantowitz and Casper, 1988). Heart rate measurement has been used in a study for determining crew rotations in the A340 (7.2.3). Heart rate measurement requires electrodes and wires, which can become uncomfortable and unstable in long duration recordings. A wristband method of measurement has been reported (Graeber, 1988). Heart rate measurement, in conjunction with other variables, has been used to asses aircrew fatigue and the effects of time zone shift. The changes in heart rate with workload are considered later (7.7.1).

7.4 Skin conductance

Skin resistance has been used extensively in psychological research and in 'lie' detection. This latter use has been controversial. The skin sympathetic nerves, part of the peripheral nervous system, control skin conductance (the inverse of resistance and the terminology most frequently used at present) by regulating sweat production. Skin conductance is primarily altered during thermoregulation and during changes in arousal. The sympathetic nerves to the skin regulate temperature by altering skin blood flow and sweat production, temperature regulation being complex with much of the neural circuitry concentrated in the brainstem and hypothalamus (5.4.2).

Many parts of the nervous system alter their level of activation during a change in arousal, modifying various body functions in a variety of ways depending upon context and specific aspects of the arousing task (Pardo et al., 1990). The neural networks involved in manifesting arousal responses from different stimuli appear to particularly use the sympathetic nerves innervating the skin. Skin conductance measurement is vulnerable to random noise in arousal and vigilance studies, the principal source of noise being that due to temperature regulation.

7.4.1 Arousal and vigilance studies

Skin conductance generally declines during a vigilance task, but the correlation between performance and skin conductance has been weak, unstable (Davies and Krkovic, 1965; Davies and Parasuraman, 1982) and complexly related to task and subject factors (Koelega et al., 1989). It has been suggested that skin conductance measurements be divided into basal level measures and the magnitude of the conductance reaction. The basal level of conductance has been proposed to be a precise measure of effort and presents an index of the load imposed by the task (Kahneman, Tursky, Shapiro and Crider, 1969) but others have felt that phasic responses better reflect task demands and or effort (Greene, Kille and Hogan, 1979). It is unlikely that skin conductance measurements can be partitioned (Koelega et al., 1989), considering the variability due to age, personality and gender effects (Davies and Parasuraman, 1982). There are many arousal, 'orienting' response, and attentional system studies that have used skin conductance measurement (Neiss, 1988).

7.4.2 Field studies

Field studies using skin conductance have been rare, particularly with respect to those in a monitoring role, because of difficulties with continuous measurement. Some laboratory studies have discarded a proportion of their conductance measurements as being unsatisfactory on technical grounds (Tulen et al., 1989). This suggests that existing methods will not be suitable for field research. Skin conductance measurement requires the use of electrodes and wires, the recording systems rarely being stable for significant lengths of time.

7.5 Adrenalin

The sympathetic division of the autonomic nervous system is involved in adrenalin release, the neural circuitry consisting of descending pathways from the hypothalamus, through the reticular formation and into the spinal cord. Sympathetic nerve fibres travel from the cord to the adrenal medulla which releases adrenalin. This system is involved in a number of regulatory control functions including glucose homeostasis and blood pressure regulation. The adrenalin release system is used in shaping a variety of body functions during arousal changes and during stress responses (5.4.1). Adrenalin measurements, which require a blood sample, are affected by many sources of biological noise.

7.5.1 Arousal and vigilance studies

Adrenalin levels decreased during a period of continuous performance and there was a positive correlation between adrenalin concentration and performance in both vigilance and complex motor tasks (Parasuraman, 1984). In arousal studies in laboratories, there were increases in adrenalin which were relatively short lived and matched the period of subjective stress (Hjemdahl et al., 1984; Tulen et al., 1989). This contrasts with the changes in noradrenalin which were much more complex, probably reflecting important interactive effects (5.4.3, 5.4.4) from the parallel outputs of the sympathetic nervous system (Persson et al., 1989).

7.5.2 Field studies

Adrenalin and noradrenalin have been measured in a variety of workplace situations in stress research studies (Balick and Herd, 1986), but not in studies directed at those in a monitoring role.

7.6 Others

Respiration rate is an inconsistent marker of the vigilance process (Parasuraman, 1984). Because respiration rate is so slaved to other functions (7.1.1), variability during most tasks is marked. Respiration rate becomes patterned during quiet wakefulness and there are characteristic rhythms in different sleep states. Respiration rate has not been used as a measure of low vigilance or low arousal.

The electrooculogram (EOG) has been used to detect eye closure. The EOG has a role in decreasing systematic noise effects in EEG recordings (7.1.1). Its use in monitoring and alerting systems is considered elsewhere (7.7.2, 8.2.2).

Body movement and body posture have been used in measuring the level of alertness. Neck flexion, head nodding, handgrip, foot position and many others have been measured with a variety of devices. Movement and posture depend upon central nervous system motor drives. The usefulness of posture and movement measures in detecting vigilance decrement depends upon the amount of decrement that occurs before the motor drive changes. These measures are described later (8.2).

7.7 Workload, effort, vigilance and arousal

Some types of measurement have been used frequently for assessing other entities besides vigilance level. In laboratory studies, heart rate variability has been used to indicate the level of arousal, stress or workload. Arousal, stress, workload and vigilance are not the same as has been repeatedly emphasised (4.2, 5.5, 6.2). The use of one measure in assessing distinctly different processes suggests that the measure is relatively non-specific.

Heart rate variability or skin conductance have been used to measure arousal. Unfortunately, most studies have shown little concern about the different roles subserved by the cardiac parasympathetic nervous system (heart rate variability) and the skin sympathetic nervous system (skin conductance). It is doubtful that two such distinct (7.3, 7.4) neural drive systems should be used interchangeably for measuring complex entities like arousal. The use of very different measures suggests that the present ways of measuring may be very imprecise and that the approach to measurement lacks rigour. Under laboratory conditions, tasks and the environment can be controlled such that a variety of measures may indicate something about the level of either arousal, effort, workload or vigilance. However, the neurobiological state is far less controllable outside the laboratory.

The problem of imprecision in measurement is much worse in reality because other entities such as workload and effort have significant effects on entities like heart rate variability and skin conductance. Thus, vigilance measurement by heart rate variability or changes in conductance would appear to be pointless for all work role applications because under different circumstances workload, effort and even fatigue will be the driving force, while at other times the vigilance decrement will dominate. There have been two responses to this problem. Either the method of measurement is further refined or other measures are added.

7.7.1 Single measure refinement

One of the commonly used measures of vigilance level, heart rate variability, has been used extensively in cockpits as a measure of workload. The way heart rate variability has been used provides an excellent example of refining a single measure.

Cockpit Monitoring and Alerting Systems

Airbus Industries have made an attempt to understand factors affecting automation of the man-machine interface, using a statistical workload calculation model (Speyer, Monteil, Blomberg and Fouillot, 1991). The model, which is based on heart rate variability, aircraft performance and situational information, has been used to relate workload patterns to level of automation, error awareness and error severity. A variety of heart rate variability indices are measured, four from the flying pilot and seven from the non-flying pilot. These are combined with ten aircraft performance measures and six flight status variables in a set of relationships which provide a predicted measure of workload. This predicted measure has been compared with an estimate of workload using the Airbus Workload Scale determined by observers and pilots during certification operations.

There is a moderate correlation between the two types of workload estimation. Higher pilot workload ratings correspond with higher heart rate and lower heart rate variability, while lower ratings are associated with lower heart rate and higher variability. In the context of highly automated cockpits, Speyer et al. (1991) have admitted that the heart rate measures may be indicating 'mental and emotional stress' rather than physical workload. Whatever is being measured, the Airbus index has dropped to low to very low levels during a lengthy flight.

The recent accidents (3.4.2) with the latest Airbus product, the A320, are cause for concern. It is possible that there is a problem with the principles and methods underlying the Airbus approach. Wide consultation with human factors experts and an enlightened approach to cockpit research appears to be associated with one result that was to be avoided, the peripheralisation of pilots such that complacency and inadequate situational awareness have become dangerously evident (3.4.2). It is a sobering thought that structural fatigue resulted in three accidents to the Comet in the 1950's and the destruction of this aircraft as a commercially viable entity. The three accidents involving the A320 have not resulted in commercial consequences so far. The duplication of the A320 man-machine interface in the A330 and the A340, both due in service in 1993, will reveal much. The long range capability of the A330 and A340 may exacerbate the peripheralisation problem.

It does not appear that the refinement of a single measure has optimised the level of 'workload', 'effort' or 'mental and emotional stress'. It is possible that the single measure has provided an index of an unknown human behavioural parameter.

7.7.2 Multiple measures

It is unknown whether the manipulation of heart rate variability measures (7.7.1) has been deemed unsatisfactory by the Airbus researchers. This is possible in the light of the latest attempts to use different measurements in developing the crew rotation schedules for the A340 (7.2.3). The most likely explanation is that the heart rate variability index (Speyer et al., 1991) was intended to measure high workloads while the studies in the A340 have been aimed at measuring low workload, or vigilance, or both.

As previously described (7.2.3), measures, including the EEG, EOG, heart rate and wrist movement, are being recorded in line operations along with observer measurements. It appears as if Cabon et al. (1991) have only combined the EEG and EOG measurements for detecting low vigilance levels. Their method of analysis may be missing a significant portion of the information contained in the data they have collected (9.3.1, 9.3.2).

Another study has attempted to provide a dynamic measure of vigilance in a work place setting. Fakhar et al. (1992) have combined six measures, two from the vehicle and four physiological measures, including an EEG ratio, heart rate variability, EOG and head angle into a vigilance state index. This approach has considerable potential (9.3.2).

7.8 Summary

Despite the importance of the vigilance process, our understanding of it is simplistic, our theories inadequate and our measurements inaccurate. There is no single physiological measure that is a reliable marker of the level of vigilance in laboratory and field studies.

Some reasons for this unproductive state of affairs are obvious. One major cause is the traditional character of the approach used to understand human nervous system function. This approach accredits focal areas of the nervous system with specific roles, a style that founders in the face of neural systems which have a widespread anatomical distribution, have 'plastic' properties, and which modulate at a myriad of sites. Another reason for the present impasse has been the failure to adopt a multi-disciplinary approach to understanding the vigilance process, an essential requirement for neural systems of such

Cockpit Monitoring and Alerting Systems

complexity and broad influence.

It is not clear that human behavioural measures such as vigilance, arousal, effort and workload can be adequately discriminated at present. While much attention has been given to distinguishing these entities, the importance of the specificity of the task, the neurobiological state during measurement, and the distant sampling of neural network function, together suggest that we do not have the luxury of making distinctions, even coarse ones. At best, prior measurement has managed to sample neural outputs, which in the context of the time and place of measurement, can be interpreted in a particular way. Whether vigilance level, arousal, or effort is being measured may be relatively unimportant, as the pattern of nervous system activation may indicate the level of engagement of a diffuse and relevant neural network.

8 Human alerting systems

Human alerting systems have had a chequered history. While it has been widely acknowledged that humans need assistance coping with vigilance decrement, the various systems have yielded few long term benefits, there being no consensus as to the optimal technology. Various types of alerting system technology will be described, a classification discussed, and the advantages and disadvantages of the different approaches considered.

8.1 Historical perspective

There have been many attempts to make alerting systems, the available technology being the major determinant of each system's attributes. Lindberg's removal of the fuselage windows of his aircraft was an admission that sustaining attention was a problem, exposure to the elements being an appropriate design change for that time (Lindberg, 1953). Lindberg is reputed to have tied a spanner to his wrist as an additional alerting system, the sudden downward drag of the spanner, as his hands slipped from the control column, acting as an alerting event. This method of signalling drowsiness is sophisticated and physiologically sound because joint movement promotes arousal.

The demand for effective monitoring in the second world war revealed how deficient humans were in the monitoring role. In a variety of situations involving lookouts and radar screen operators, vigilance decrement was found to be a major problem. Job redesign with duty breaks during the first 30 minutes of monitoring were found to be useful but alertness indicators were also considered. Early research focused on measurement of the amount of muscle contraction in the neck (Kennedy and Travis, 1947). Since then vigilance has been measured in a number of ways (7.2-7.7) but attempts to incorporate these measurements into practical alertness assisting systems have met with little success.

Cockpit Monitoring and Alerting Systems

8.2 Road transport alerting systems

A variety of alerting systems have been proposed for road vehicles. Alerting systems have been divided into two types, those that detect changes in driving performance and those that signal changes in the vigilance level of the driver (Tarrière, Hartemann, Sfez, Chaput and Petit-Poilvert, 1988; Vallet, 1991). There is a third category which can be distinguished, namely those that measure vigilance performance, rather than the level of vigilance.

8.2.1 Driving performance measurement

These systems detect the pattern of steering wheel movements. The proposition underlying their use is that the pattern changes as vigilance level declines. EEG studies have suggested that as vigilance declines, steering wheel movements become less frequent but larger and there are significant changes in vehicle tracking (Chaput, Petit, Planque and Tarrière, 1991).

Driver alertness aid Developed in the 1970's, this device detected steering wheel movements by contacts which signalled when the steering wheel had deviated more than two degrees from the mid-line position. A reference period was used and an alarm sounded if there were too few contact signals. An orange light came on if there were too many. This device was not successful, corners proving confusing and the three minute reference period too short. The frequency of steering wheel movements, rather than the character of steering wheel movements, does not provide a measure of attention level (Tarrière et al., 1988).

Safety drive adviser This device was incorporated in the Nissan Bluebird as sold in Japan from 1985 to 1986. A reference period was used to record the amplitude and speed of steering wheel movements, an alarm sounding if subsequent changes exceeded certain threshold values. Despite being installed in a significant number of vehicles, there has been no report of the device's efficacy (Tarrière et al., 1988). This device could cope with the change in steering wheel patterns associated with cornering.

Vigilynx This is a contrast sensor able to detect the change from tarmac

Human Alerting Systems

to road edge or grass. It can also detect the crossing of a continuous or broken white line, but it is possible that it misses the onset of the vigilance decrement significantly. A variety of similar systems are under development by different major car manufacturers.

8.2.2 Vigilance state

There are many devices designed to detect diminished vigilance, there being more than 200 patents in this category in the USA. Some of these have been used, although their performance claims have not been supported. Some have a simplistic splendour.

Dozer's alarm This device detects head nodding. The proposition underlying its design is that head nodding signals vigilance decline. Head nodding is complex, being associated with fatigue but usually only commencing once there has been substantial loss of attentiveness (Haworth and Vulcan, 1991). An on-road study using multiple cameras monitoring head and body position has suggested that there may be complex postural changes with driving which better reflect vigilance level than head nodding (Fakhar, Vallet, Olivier and Baez, 1991).

The dozer's alarm is commercially available and consists of a hooked ear piece which contains a battery, an alarm and an angular rotation detector. Head nod beyond a certain angle makes the alarm sound. The device has been tested. Sometimes the device signals driver fatigue, but it does not alter the ability of the driver to drive for longer or to drive with less performance decrement (Haworth and Vulcan, 1991). The device has not been considered by drivers to be an impediment.

Variants of this device have been reported anecdotally. The simplest version has the nodding head coming into direct contact with a bell. Slightly more sophisticated is the tie from a helmet to the klaxon switch on the roof above, a head nod resulting in a horn blast. The device is reputedly fitted to a large truck, but its effectiveness can be questioned, because observers know when this truck is nearby from its repeated horn blasts.

Dormalert This device measures skin conductance, the underlying proposition being that skin conductance reflects the level of vigilance (7.4). While skin conductance has been a popular measure of arousal in the laboratory, this may not be the same as measuring vigilance (4.2,

5.5, 6.2, 7.7). There is disagreement about how skin conductance measurements should be analysed (7.4.1). Conductance measurement is plagued by random and systematic noise related to fluctuations in the thermoregulatory drive (7.4) and stability has been a problem in field studies (7.4.2). In the context of road driving, thermal effects might dominate measurements.

The Dormalert has ring electrodes for a finger and a toe, wires and a processing unit, which can fit in a pocket. A calibration period of 10 to 15 minutes is used to determine a detection threshold which reduces spurious alarms. The device has been tested but is not obviously effective (Tarrière et al., 1988). The period of calibration and the setting of the threshold have no guidelines. The electrodes and wires provide a significant disincentive for routine use.

Onguard eye closure monitor The duration of eye closure is related to increasing drowsiness, the number of long duration closures being related to fatigue, the time spent driving, tracking errors and crash risk (Haworth and Vulcan, 1991). Extended eye closure directly affects the process of vigilance.

The onguard device consists of a small infra-red sender and sensing unit, processor, batteries, switch and alarm all mounted in a small module attached to a spectacle frame. The amount of reflected infra-red is reduced when the eyelid is closed, closures of longer than half a second causing the alarm to sound. The device has been tested. Sometimes the device signals fatigue, but it does not alter the ability of the driver to drive for longer or to drive with less performance decrement (Haworth and Vulcan, 1991). Good fitting spectacle frames are essential.

Electroencephalogram and electrooculogram Present research is focused on the properties of the EEG and the EOG which can usefully be combined in an algorithm to give a single vigilance measure. The algorithm is being refined for the detection of reduced vigilance (Berrichi, Tibergé and Arbus, 1991). While this approach is likely to be limited by the problems associated with EEG and EOG measurement of alertness (7.2.1, 7.2.2), the methods appear to be in advance of those employed in the latest Airbus studies (7.2.3, 7.7.2). In a road system, recording electrodes would be carried in a specialised set of spectacle frames. Driver acceptance and artefact free recording may be problems.

Magneto encephalogram Instead of recording small currents in scalp skin, this research program is measuring attentiveness by recording the minute fluctuations in the magnetic field associated with cortical currents. Skin electrodes are not required. This approach has the same limitations as those that use EEG measurements for assessing the level of vigilance (7.2.1). Some doubt that the magnetic field fluxes are measurable in that they are very much less than the earth's magnetic field and may not be recordable in a moving vehicle (Vallet, 1991).

8.2.3 Vigilance performance measurement

Some systems measure vigilance performance rather than the level of vigilance. They do not measure driver performance.

Roadguard This device uses the observation that subsidiary task reaction time increases with the hours of continuous driving. A number of studies have shown that the reaction time of tasks involved in driving, such as brake reaction time, do not provide a satisfactory measure of fatigue. In contrast, subsidiary task reaction time changes with fatigue, only the driver with the double task of driving and managing a subsidiary reaction time task showing an effect (Haworth and Vulcan, 1991).

The roadguard device detects fatigue by monitoring reaction time using a visual signal and a reaction time clock. An electronic circuit is activated as the vehicle is put into top gear, part of the circuit being a timer which stops at random periods after 4 to 14 seconds. When the timer stops, a light on the dashboard comes on. Failure to switch the light off by contacts on the steering wheel or a foot switch results in a buzzer sounding. Switching the light off resets the timer.

The roadguard device has been able to detect the onset of fatigue but has not resulted in longer driving before falling asleep. There has been no alteration in tracking performance when it has been in action. The device in its present form can be circumvented by pressing the contacts about every four seconds and ignoring the lamp. An automatic pressing mode (8.5.3) can be present (Haworth and Vulcan, 1991). It is possible that the need to respond to the Roadguard device interferes with driving performance.

Artificial skin This device detects the pressure exerted by the driver on

Cockpit Monitoring and Alerting Systems

various controls, the underlying proposition being that multi-sensorial measurement of effort will indicate changes in vigilance. Pressure is detected by multiple sensors. However, pressure applied to the steering wheel is not correlated with changes in vigilance (Tarrière et al., 1988), nor does pressure measurement equate to performance measurement. In many respects, this device is similar to the 'dead mans' handle (Vallet, 1991).

8.3 Rail transport alerting systems

Rail transport systems have a variety of methods for measuring driver vigilance and coping with the consequences of inadequate attentiveness. Some systems are external to the train itself. Thus, trip switches beside the line stop trains if they go though red lights and trains can be derailed if required.

The 'dead man's handle' comes in various forms, its primary role being to detect driver incapacitation. The 'handle' may be part of the throttle control circuit or may only be connected to the braking system. The handle can also be a pedal. The vulnerability of this system has been manipulated in the world of fiction and film, a basic plot having three key elements; a very inattentive or dead driver, a runaway train, and debris, tool boxes or other objects jamming the handle.

8.3.1 Secondary task systems

Most modern locomotives have the facilities for measuring vigilance performance using a subsidiary reaction time task. However, the way these tasks are used suggests that the tracking of fatigue is not their primary role. Rather, the system is used solely to assert that someone in the driving compartment is awake enough to respond to a buzzer, light or soft bell. These systems are identical to the roadguard device (8.2.3) but the random periods are much longer. However, the random period is being reduced as the potential speed of locomotives increase.

SNCF The SNCF system uses a constant position, constant pressure foot pedal that does not set the speed of the train. Periodically (approximately every minute), a soft sound prompts the driver to demonstrate that vigilance levels are okay. The demonstration requires relaxing the foot control within 2.5 seconds of the prompt, a failure to respond leading to a klaxon alarm which if not turned off in another 2.5 seconds initiates an automatic stopping sequence (Masson, 1991).

SIFA The German Federal Railway has installed a paced secondary motor task, called the train function safety circuit (SIFA). This monitors the train driver's fitness for the monitoring role. The normal function of the train is ensured providing the SIFA control is depressed for up to 35 seconds and released for no longer than 5 seconds. Lights, followed by horns and emergency breaking start from the 30 second mark of continuous lever pressure. The key feature of this device is that it is not connected in any way to the speed control of the locomotive and exists purely as a device to maintain vigilance. The device's effectiveness has been questioned (8.5.3) and recently tested (Peter et al., 1990b).

8.4 Aircraft alerting systems

A crew performance system (CPS) has been fitted to the Boeing 747-400. As far as is known, there is no other person performance, vigilance performance or vigilance measurement system in commercial aircraft.

8.4.1 Historical aspects

It is likely that the CPS in the Boeing 747-400 was developed at the request of an individual airline. The development phase occurred in the mid 1980's. The airline has insisted on remaining anonymous[2], perhaps because of the publicity surrounding incidents in which pilots have fallen asleep in cockpits (Beech, 1991). In satisfying certification requirements, Boeing must have carried out considerable integration and testing, probably in association with NASA.

8.4.2 Probable characteristics

It has not been possible to see the specification of the CPS fitted to some Boeing 747-400s, nor has it been possible to examine airline documentation. Boeing considers that the propriety issue[3] is significant.

---

[2] Letter from Editor of Flight International.
[3] Personal communication, Boeing 1992.

Cockpit Monitoring and Alerting Systems

The Boeing CPS is likely to have many of the following features.

The system is entirely software based and has no strap on sensors, dedicated switches, buttons or warning lights. The CPS functions as if all cockpit switches, levers and other controls have parallel connections, such that the use of any cockpit item can be logged. If there is no input into the CPS over a certain time period, one of the visual display units asks for an input. An absence of a response results in an alarm sounding in the cockpit. The system has features of a performance measurement system (8.2.1), but is used purely to assert that someone is awake enough to respond to a visual prompt.

It is not known if the alarm sounds in the adjacent crew rest area. The time period is likely to be in the order of twenty minutes, but is obviously alterable. It is unlikely that the CPS is an option controlled by the pilot, the choice probably residing with flight operations management. The fate of the information that such a system can obtain is unknown. It would be feasible for the CPS to track the number of visual prompts and the resulting response times.

8.5 Alerting system classification

There have been few attempts to classify alerting systems. A taxonomy of countermeasures to loss of alertness in truck drivers has been formulated (Mackie and Wylie, 1991). The following classification for cockpit alerting systems does not consider the alertness maintainers that are operator controlled, such as auditory stimulation, mental games, environment control, stimulants and rest stops. Nor does it consider external controllers of alertness such as shift rotation, working hours limits, obligatory rest stops and training systems.

A simple classification useful for commercial cockpits divides alerting assisting systems into three major divisions; those which measure person performance, those which measure vigilance performance, and those which measure vigilance level. All of these can be linked to systems which manipulate crew performance, vigilance performance or vigilance level. Manipulation can be by lights, sounds or visual and verbal prompts. Manipulation in the future may include altering the style and type of automation, change in use of human resources and the use of chemicals. The effectiveness and consequences of the various systems will be considered.

8.5.1 Alerting system effectiveness

There are a variety of effectiveness criteria (Mackie and Wylie, 1991), including reliability of measurement, sensitivity to operator impairment, false alarm rate, adaptation to environmental conditions, obtrusiveness, complexity of analysis, duration of effect, acceptability to operator and relative cost. These effectiveness criteria have been applied in various ways to truck (Mackie and Wylie, 1991) and car alerting systems (Haworth and Vulcan, 1991).

At this point in time, there is no alertness assisting device in any transport system which has had its effectiveness publicly demonstrated. Factors altering effectiveness may include (Mackie and Wylie, 1991):

1. Person performance systems are vulnerable to individual idiosyncrasies and to environmental changes.

2. Vigilance performance systems interpose themselves between the task and the operator, the diversion of operator attention causing rejection by the operator.

3. Vigilance level systems are often intrusive, the interference impairing long term acceptability.

Prior considerations of alerting system failure have been thorough but have been focused on systems involved with procedural tasks rather than on monitoring. The failure of all these systems in procedural roles bodes ill for their use in monitoring roles. The general failure of all alerting assisting systems suggests that there may be other factors which are far more potent than previously thought. Other factors may include gaol alignment, role ambiguity, peripheralisation, site of control and complacency.

8.5.2 Effectiveness of person performance system in monitoring role

Person performance systems measure human activity. When activity levels shift outside arbitrary standards, a warning is given or an alarm sounded. The Driver Alertness Aid and the Safety Drive Adviser are road examples (8.2.1), while the Boeing CPS is an aviation example (8.4). Effectiveness has not been demonstrated.

Cockpit Monitoring and Alerting Systems

The key advantage of person performance systems is simplicity and the ease with which performance data can be obtained without having to instrument humans. No test system intervenes between the operator and the task. There are disadvantages related to system goals and effects. Consequences from their use suggest that they have no place in cockpits.

Goals Performance becomes more tenuously related to vigilance or arousal as the procedural content of the task decreases. In the extreme case of pure monitoring, performance measurement systems confuse the absence of evidence with the evidence of absence. These types of devices can never be alerting systems in monitoring situations where procedural activity is very low.

Effects Person performance systems do not provide means for altering performance. They give no information as to why performance is unsatisfactory. Thus, the Boeing CPS penalises pilots who maintain alertness but do not touch the controls. If the time period is shortened, the degrading situation can occur of the pilot having to repeatedly press some innocent button to keep the system silent.

Consequences Person performance systems promote peripheralisation in automated cockpits, because the alerting system demands responses that have nothing to do with the task. This misalignment of goals is likely to lead to either automatic behaviour (8.5.3) or creative disablement (8.5.5). The Boeing CPS is likely to increase role ambiguity and vigilance stress because it repeatedly reminds pilots of their peripheralised status. There is no evidence that any of these systems are effective, nor have the road examples been further developed. Their development and certification costs must have been significant.

8.5.3 Effectiveness of vigilance performance system in monitoring role

Vigilance performance systems measure human performance at a vigilance task. It is possible that they can indicate change in vigilance processes. The Roadguard device (8.2.3), the SIFA system (8.3.1) and the Psychomotor Vigilance Task (8.6.1) are examples. Effectiveness in detecting fatigue in a procedural task has been demonstrated (8.2.3).

The key disadvantage that has been well recognised is that vigilance performance systems interpose themselves between the operator and the primary task (8.5.1). In the SIFA system, the paced secondary task has lead to some unwanted effects.

In a laboratory study on the SIFA system (8.3.1), experienced train drivers were able to operate the vigilance maintaining device more effectively than a group of untrained truck drivers. However, the experienced group regularly reached the stage of light sleep, showing distinct decreases in alertness compared with the inexperienced group (Peter et al., 1990b). The experienced train drivers were able to avoid triggering the emergency braking system, despite hypoarousal. The same phenomenon has been observed by others, road drivers developing a rhythmic and semiautomatic way of operating secondary task systems with low levels of vigilance (Mackie and Wylie, 1991). The likelihood that the primary task may be affected detrimentally by automatic behaviour would be a major disadvantage, the whole point of the secondary task system being lost. The secondary task system may be doubly dangerous, because of the complacency it produces amongst those whose task it is to regulate public safety.

Other disadvantages of vigilance performance systems used in a predominantly non-procedural environment relate to goals and effects. The consequences suggest that vigilance performance systems have no place in cockpits.

Goals While it is likely that vigilance performance systems provide a measurement of the vigilance process, it is possible that this process is not relevant to the vigilance task required. The specificity of processes in the vigilance control system and in the linkages between vigilance networks (5.3) suggest that systems like the SIFA circuit might track vigilance performance only during a sensorimotor task. At present, the SIFA circuit is not used as a tracking device. Similarly, Roadguard might only signal vigilance performance with respect to visual detection of dashboard lights. Neither outcome is what is desired. The lack of alignment between the vigilant behaviour required for the primary task and that required by the vigilance performance system may prove to be a major problem.

Effects The automatic behaviour phenomenon is well described and at odds with the design intent of vigilance performance systems. Other unwanted effects include creative disablement (8.5.5). Vigilance performance systems demand activity unrelated to task, without giving

control to the operator or the resources to manage performance shortfall.
Consequences These are similar to those for the person performance system (8.5.1) and include further peripheralisation, role ambiguity and vigilance stress. There have been no reports that vigilance performance systems have useful effects outside laboratories.

8.5.4 Effectiveness of vigilance level systems in monitoring role

Vigilance level systems measure human variables which are purported to indicate the level of sustained attention. When this variable shifts outside certain standards, a warning is given. The Dormalert, the Onguard eye closure monitor and the Dozer's Alarm (8.2.2) are road examples which have all proved to be ineffective.

The goal of measuring vigilance level is a worthy one, because these measures focus directly on the processes underlying monitoring. Their key disadvantage is their intrusiveness. They have relatively few effects or consequences. The ineffectiveness of the devices produced so far is further evidence of the pointlessness of using single physiological measures in determining vigilance level (7.7.1-7.7.2). There are other causes for ineffectiveness related to alarm timing with respect to the vigilance decrement and the inappropriateness of the measures used.

Timing of alert Head nodding and increases in eye blink duration may occur after significant vigilance decrement has already occurred. The sequence of changes in different physiological parameters with vigilance decrement is not clear, but some changes probably occur consistently later than others. The reproducibility of the sequences in the work place for one individual for a single task, for the same individual for a different task and for different individuals has not been described.

Appropriateness of measure Single measures in themselves are far from ideal (7.7). Attempts to use the EEG in the workplace have met with little success (Davies and Parasuraman, 1982). Prior studies related to air traffic control systems avoided the need for alertness displays and continuous recording. Even with EEG feedback and using the most advanced methods of recording, it is likely that EEG measures in isolation will be disappointing (7.2.2-7.2.3) Other single alertness measures have been discussed (7.2-7.6).

8.5.5 Creative disablement

Automatic behaviour is a definite impediment in vigilance performance systems (8.5.3). Creative disablement is equally as difficult to manage. Creative disablement encompasses all those activities that humans use to bypass systems that control behaviour.

There are some ingenious examples in cockpits. Pilots flying the Boeing 757 can have trouble with programming the flight management system (FMS) for an early descent. Traffic and weather conditions can make this descent desirable but the FMS can be resistant to alterations in the flight plan. Pilots remove this restriction by programming in a falsely high tailwind, the original descent point being reached prematurely by using misinformation. This type of activity is quite distinct to automatic behaviour.

The underlying factors which result in creative disablement are not well understood. They may include job dissatisfaction, poor man-machine interface relationships, goal dissonance, the challenge and problems with site of control.

8.6 Cockpit systems under development

Two approaches which may assist pilots in their monitoring role are under development at present. The final configuration of each system is unknown and their effectiveness has not been reported.

8.6.1 Psychomotor vigilance task

The Psychomotor Vigilance Task (PVT) is designed to indicate fatigue in long-haul operations (Dinges and Graeber, 1989). The PVT is a short duration (10 min) vigilance task which is able to detect monitoring lapses and the decline in optimum response capacity. It has been proposed that the PVT be employed as a secondary task system in long distance cockpits (Dinges and Graeber, 1989).

The PVT asks those in a monitoring role to periodically assert their right to continue monitoring. In this role, it is beset with a number of fundamental problems. The ingenuity of humans in bypassing the SIFA circuits in German trains (8.5.3) suggests that secondary task systems may not deliver the sustained level of attention expected (Peter et al.,

1990b). The PVT also interferes with the monitoring role, the duration of the secondary task being long enough to significantly interfere with the monitoring task. Finally, repeated assessment by a secondary task system reinforces role ambiguity and in long duration flight is likely to have a significant stress effect.

8.6.2 The pilot's associate

The Pilot's Associate or electronic copilot is an interface between the pilot and the aircraft which recognises all the pilots who operate a particular plane. Each pilot's knowledge and experience are known as the Associate prompts, advises, reminds and warns its human companion. Such a system is being planned for use in the Advance Tactical Fighter in the mid 1990's (Warwick, 1989). The assumption behind the Pilot's Associate is that intelligent systems can bypass the need for knowing about the person directly by dint of extraordinary memory and continuous comparison of past and present performance. This approach avoids interrupting the monitoring function (cf. PVT, 8.6.1), but is as vulnerable to human manipulation. Its presence will promote role ambiguity. Although it has the potential to be useful, in isolation the Pilot's Associate does not assess the level of vigilance and is only a sophisticated version of the Boeing CPS (8.4, 8.5.2).

8.7 Summary

It is apparent that those who are responsible for designing the pilots' future workplace are concerned about the ability of humans to sustain attention as are others involved with other transport systems. It is of concern that those responsible for monitoring situations with a far greater potential for harming huge numbers of people appear so disinterested in the issues surrounding the monitoring role.

The attempts that have been made to assist humans in the monitoring role have been uniformly ineffective and suffer from multiple flaws. Performance measurement systems appear to be the most primitive, being no more than an electronic variant of a 'time and motion' study. Continuous vetting of human endeavour has never, and will never, produce a quality service, contribution, or product. Vigilance performance systems are also badly flawed and ineffective. The flaws vary from interposing functions between the operator and the task, to the repeated reminders of operator inability to carry out the monitoring function. In the context of highly automated systems, performance measurement and vigilance performance systems will amplify peripheralisation effects by promoting role ambiguity and vigilance stress. Vigilance level systems are equally ineffective, although the issues are probably different. Their ineffectiveness reflects their intrusiveness and a very limited appreciation and understanding of the vigilance process.

9 The ideal alerting system

Previous sections have detailed some of the problems bedevilling the construction of human alerting systems. These problems are reviewed, the characteristics of an ideal alerting system and the arguments for multiple measures of vigilance level are outlined, and some practical features of an alerting system are discussed.

9.1 Established objections

A number of problem areas require further consideration before a cockpit alerting system can be constructed. Problems include:

1. There is deep scepticism amongst those who study vigilance concerning the practical application of vigilance research work (5.1). The profusion of vigilance theories, the encapsulated and narrow range of vigilance tasks examined in vigilance research, and the failure of vigilance theories to provide useful solutions together suggest that the transfer of research findings from laboratory to cockpit should be carried out very cautiously.

2. The same scepticism, doubt and inapplicability applies to much of the work on arousal (5.2). There is confusion about the concept of arousal, its nervous system substrate, its place in concepts like workload, effort, fatigue and vigilance, and the applicability of most arousal research. Any application must accept that arousal may only have meaning for specific neurobiological states.

3. Prior approaches to alerting system design have failed to give operators choice with respect to the assistance they receive. Alerting systems which remain outside the operator's control promote automatic behaviour or creative disablement (8.5.5).

Cockpit Monitoring and Alerting Systems

4   Alerting systems that measure person performance (8.5.2) or vigilance performance (8.5.3) do not appear to work. Measurement of vigilance level (8.5.4) has not worked either but for different reasons. Inadequate understanding of vigilance processes and the use of single measures have been the problems.

5   Physiological measurements of vigilance level have been useful in research settings but are often impractical in the workplace. They are intrusive and become 'noisy' in uncontrolled environments.

6   The links between arousal and vigilance and the effects of cognitive stressors emphasise how fragile and vulnerable the monitoring process is. It is unlikely that a behavioural modification program, a human selection technique or a single piece of technology can provide a solution in isolation.

It comes as no surprise that alerting assistance systems have been dismissed. It is likely that all existing approaches are flawed to some extent.

9.2 Ideal system characteristics

The lack of understanding of the vigilance process, the lack of agreement on the concept of arousal, the confusion surrounding the arousal-performance relationship and the limited utility of vigilance research findings might suggest that the development of vigilance assisting systems is a questionable activity. However, prior efforts and an evolving conceptual understanding allows definition of essential characteristics of an ideal alerting system. These include:

1   The alerting system must measure the vigilance of the operator.

2   The alerting system must not impede, intrude or interfere in any way with the operator.

3   The alerting system must be tailored to each operator, the operator retaining control over the degree of assistance offered.

The Ideal Alerting System

4   Access to the alerting system must be controlled by the operator.

5   The alerting system must be robust and by its design, must remove the desire of those monitoring to want to manipulate it. There must be a synergy between the design goal of the alerting system and the monitoring goal of the operator.

6   The alerting system must be designed such that it is suitable for everyday use by a multitude of personality types across different cultures. The system must, in its measurement process and its means of altering alertness, cause no harm to health in the short or the long term.

9.3 Vigilance measurement

The alerting system must measure the vigilance level of the operator. The use of single measures and the failure to accommodate the multitude of neuro-affective states have been major barriers. Vigilance processes reside in intertwined neural networks (5.3) and multiple measures are required. There are many arguments for multiple measures.

9.3.1 The case for simultaneous physiological measures

There are three lines of evidence which suggest that the simultaneous measurement of multiple physiological outputs is the direction that measuring vigilance level should take.

Single measure studies A number of studies of arousal and vigilance have shown that measurement of a single physiological entity is of doubtful benefit (7.2-7.6, 7.7.1-7.7.2, 8.2.2). Single outputs of various types can signal aspects of human activity in the cockpit (Speyer et al., 1991), although each, no matter how processed it is, does not provide a measure of the level of vigilance. Recent fatigue and time-zone shift studies, using multiple measures (Graeber, 1988; Cabon et al., 1991), point the way measurement processes are moving. Studies in road vehicles which measure multiple physiological entities and combine them into a vigilance index (Fakhar et al., 1992) predict the likely way multiple measures will be used.

Multiple measure studies A number of studies (Schnore, 1959; Elliott, 1964; Neiss, 1988) have suggested that multiple measures allow for the individual patterning that each psychobiological state produces in each individual (Neiss, 1988). The recent attempts to measure vigilance using two or more physiological measures in road vehicles (Berrichi et al., 1991; Fakhar et al., 1991; 1992) demonstrate that it is possible to manage a range of implementation issues and obtain useful information. The studies in the cockpit (Graeber, 1988; Cabon et al., 1991) have used very similar measures in assessing the effects of time zone shift and fatigue. The use of similar measures for assessing vigilance or fatigue suggest that assessment of the level of activity in a particular neural network will depend upon the processing paradigm.

Although there is no clear agreement on the number of physiological outputs (9.3.2) that should be measured, multiple measurement allows an assessment of the pattern of neural network activity, activity which is the orchestrated end affect from vigilance and arousal neural networks.

Clinical studies In clinical medicine the problem associated with diagnosing dysfunction of the autonomic nervous system is very similar to the problem of trying to measure vigilance level. Single measures of brain stem function correlate poorly with the state of known autonomic nervous system disease. However, when multiple simultaneous tests are used to measure the function of different parts of the autonomic nervous system, their diagnostic specificity and sensitivity is significantly improved (McLeod and Tuck, 1987; Ewing, 1988). It is now routine to carry out five simple screening tests of autonomic function on patients presenting for investigation. Abnormality in one is disregarded, but abnormality in two, or abnormality in one and two other borderline results correlates well with autonomic nervous system dysfunction proven by invasive methods.

Summary Neural networks have idiosyncratic and specific properties such that significant patterning of responses is to be expected. No two person's patterns will be identical across all vigilance activities and in any one individual, patterns will depend upon the mix produced by environmental and internal drives. It is not surprising that a single measure is less useful than expected, the more measures the more reliable the characterisation of that person's pattern.

9.3.2 The choice and number of measures

The measurement system must sample activity levels in a number of neural networks, particularly those frequently implicated in vigilance processes. The measurement system must also sample other outputs and physical variables which are sources of systematic and random noise (7.1.1). Thus, factors which dictate choices of variable include breadth of network sampling, specificity with respect to vigilance processes and the effect of significant modulating factors. The number of measures is dictated by the minimum required to index the vigilance level in a satisfactory manner under different circumstances. Both number and choice are likely to vary, depending upon the psychobiological state. Flexibility in information acquisition is required.

Choice Parts of the vigilance process have been imprecisely measured with some form of EEG recording (7.2.1-7.2.3). These types of measurement need to be mixed with others which focus on other aspects of vigilance network function. The closely intertwined nature of the arousal and vigilance processes, particularly in relation to the admixture of the networks in the hypothalamus provides direction (5.3-5.5). Thermoregulatory systems are tightly linked to sleep cycle systems. A measure of the thermoregulatory motor drive as well as the state of the environment allows the arousal component of the thermoregulatory drive to be determined. The adrenomedullary system is also important and should be sampled, although divorcing cardiovascular factors with the use of auxiliary measurements is difficult. Similarly, other cardiovascular variables such as heart rate, heart rate variability, blood pressure, blood pressure variability and pulse pressure are relatively peripheral to vigilance processes but may be considered as long as some measure is made of factors which profile systematic and random noise effects (isometric exercise, workload, etc.). Measures of posture and eyelid position may have a place, once the cascade of physiological changes, which occur during the vigilance decrement in the workplace, has been determined. Some of these networks may require more than one measurement device but one measurement device may provide more than one output in different psychobiological states.

Number It is unknown whether there is an optimal number of measures, some suggesting that many variables are required (Schnore, 1959;

Cockpit Monitoring and Alerting Systems

Elliott, 1964; Neiss, 1988). The recent automotive vigilance study combined four physiological measures (Fakhar et al., 1992). A more important issue is not the number, but the processes whereby the choice and hence number of measurements is known to be sufficient. Vigilance performance can be measured in the workplace, allowing the determination of the minimal number of measures, which when combined, give a useful estimate of vigilance level.

9.3.3 Measurement methods

Physiological measures in their present form are mostly unsatisfactory, because all of them interfere with the person monitoring. Electrode systems interfere in a myriad of ways and have little application for long term usage. Problems arise with wires, electrode paste, electrode position, the need to adopt certain postures and avoid others, and the need to reapply electrodes from time to time. There is no place in routine commercial operations for cockpit crew using systems which involve wiring or tethering of any sort. Helmets are not an option although a headset appears to be well tolerated.

The use of a wristband to record three physiological variables in pilots in long haul sectors (Graeber, 1988) demonstrates that the ideal of minimal to no restriction can be attained. A variety of methods allows sampling of many physiological variables without direct contact. An example, which demonstrates solutions to a number of measurement issues, is described later (9.5.2).

9.4 Operator control

Present experimental vigilance measuring systems have the operator trussed and wired, cosseted and corralled, both physically and psychologically. The level of restriction imposed by a future commercial aviation alerting system will have to be minimal (9.3.3). Unless the operator has control of the level of assistance, with its accompanying restriction, it is likely that the assistance that is on offer will be bypassed (8.5.5).

Some cockpit systems, such as modern map displays, have produced immediate synergies between the pilots' wishes and those of others with vested interests. These stakeholders include the aircraft operator, air

The Ideal Alerting System

traffic control personnel, passengers and aircraft designers. Other cockpit systems, such as the Ground Proximity Warning System (GPWS), only achieved a limited initial acceptance. Over time, the GPWS has been developed such that all now benefit from its presence.

9.4.1 Site of control

Operator control occurs at a number of levels. The design and development phase of a vigilance level measuring system requires considerable input from pilots. Optimal system characteristics will depend upon pilot participation in the design phase.

Implementation, and the development of procedures for optimising an alerting system, should reside principally in the control of flight deck personnel. Just as the development and tuning of software for fly-by-wire systems continues for long after a new aircraft flies, so the characteristics of the alerting system will be moulded at the preliminary operations level and also during line operations. This approach allows modulating factors like specific flight profiles and organisational culture to exert their effects.

In everyday use, it is essential that flexibility in configuration and function are integral and outstanding features.

9.4.2 Methods of operator control

Methodologies underlying operator control are under development. They are intimately linked with the concepts of adaptive automation and the Pilot's Associate (8.6.2). The Pilot's Associate is at present labouring under the yoke of being a performance measurement system, which is unlikely to be successful. The performance measurement approach is outdated, in that it relies on quality assurance, which fails to consider the means whereby performance might be altered. The setting of standards, while ignoring essential processes, has little future. However, it is well within the capacity of the Pilot's Associate to be a true associate and not a measurement and performance master.

A key function of the Pilot's Associate should be the processing of physiological data transmitted from the variety of sensors. This information is converted to an index, the choice of which can reside with the pilot. A suite of operator and task-specific algorithms will individualise each pilot's index and display it in a form of the pilot's choosing. How the index is used in the context of other Pilot's Associate features is open to the pilot and copilot.

The Pilot's Associate has already been assigned the task of knowing each person and their performance at various flight tasks (Warwick, 1989). This performance measurement function can be changed from the supervisory quality assurance role to that of true assistant, able to help in changing performance, providing the state of the person is known. Thus, instead of using an approach where standards and prior performance provide benchmarks, operator vigilance levels enable the pilot and the Associate to optimise the human resource together.

9.5 Measurement issues

There are three measurement issues that are critical initial problems. One issue is obtaining restraint free, non-intrusive measures of vigilance level. The second is obtaining continuous measures which provide reproducible patterns from one flight to the next and which are usable in producing a vigilance index. The third issue is controlling for random and systematic noise. A skin conductance measurement system (Figures 9.1 and 9.2) illustrates how existing systems might be changed and how these initial issues can be approached.

9.5.1 Prior approach to skin conductance measurement

Thermoregulatory systems are tightly linked to sleep cycle systems and are important for physiological measurement in a vigilance system (5.4.2, 9.3.2). A measure of the thermoregulatory motor drive as well as environmental and body temperatures allows the arousal component to be determined.

Prior studies have rarely progressed beyond the laboratory, electrodes and wires providing restraints and the electrode skin-interface being poorly stable in the long term (7.4.1-7.4.2). Two attempts to use conductance outside the laboratory have been failures, but new approaches (9.5.2) promise to bypass many prior problems (Figure 9.1).

An electrode system for detecting the skin conductance increase caused by insulin induced hypoglycaemia in diabetics (Hansen and Duck, 1983) has not been successful because the electrode-skin interface was unstable. The Dormalert (8.2.2) has proved equally unsuccessful, the

The Ideal Alerting System

A.

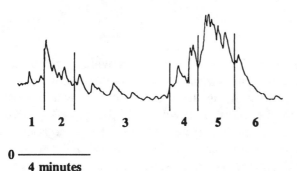

B.

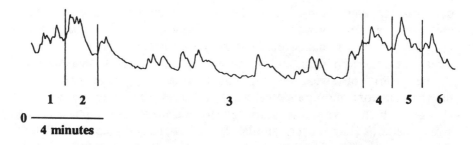

Figure 9.1 Sweat output from the left index finger during two short duration simulator flights. A Beechcraft Starship was simulated using the Microsoft Flight Simulator running on an AT microcomputer. The take off was from Meigs airfield, landing by the shortest path (A) and a longer route (B) at Chicago O'Hare airport. A twenty knot wind was blowing from the west at sea level up to 5,000ft. The flight phases are taxi (1), takeoff and climb (2), cruise (3), descent and initial approach (4), final approach and landing (5) and taxi (6). The units of sweat are arbitrary, but the same for both flights

restraint from electrodes and wires and the instability of the measurements creating problems.

9.5.2 New approaches

It is possible to measure the activity of sweat glands without electrodes by measuring the amount of sweat produced continuously. A sudorometry method relies on blowing dry nitrogen or air over a piece of skin and determining the humidity or dewpoint of the gas after it has been in contact with the skin. Such a system is free of the vagaries of electrode-skin contact, the system can be calibrated and measurement can continue for extended periods. Such a system can be restraint free, does not interfere with monitoring function, and does not interfere with flying. This type of system can be extremely sensitive depending upon the gas flow rate, the skin area and the sensor.

Electrode free recording Sweat patterns have been measured with a preliminary model during simulator sessions (Figure 9.1) in a constant temperature environment. In most flights the sweating patterns have been similar, there being increments in sweat output at take-off and in all phases of the approach and landing. During the cruise phase, sweat output settles and remains relatively stable. In some extended flights the sweat output becomes flat and featureless during the cruise phase.

The profile of sweat output, which is the output of part of the thermoregulatory and arousal networks, has been similar to the heart rate changes which have been described for different flight phases[4]. The sweat patterns have been similar in many respects to the patterns described for pilot workload. These have also been derived from recordings of heart rate (Speyer et al., 1991). Under conditions in which the thermoregulatory drive is constant, the sudorometry measure is giving a measure of arousal. The neural networks responsible for skin sweating and for heart rate control are quite distinct.

Random noise effects Using the sudorometer system described above (Figure 9.1), different random noise effects can be appreciated. The thermoregulatory drive has a significant impact (Figure 9.2 (A)), as expected, and could be managed by intelligent processing, providing the

---

[4] Figure 6.3 Kantowitz and Casper 1988

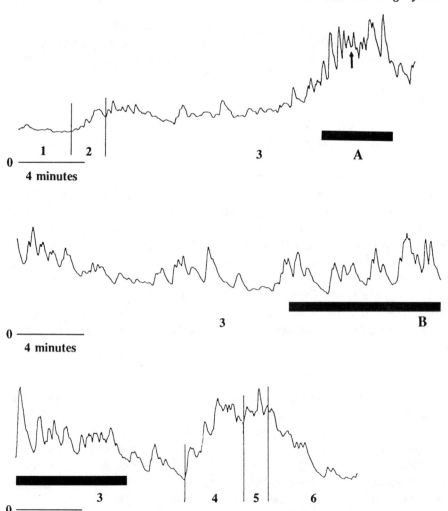

Figure 9.2 Sweat output from the left index finger while flying a simulated Beechcraft Starship south from Chicago O'Hare airport and return. Flight phases, simulator details and sweat details are identical to Figure 9.1. Soon after takeoff (A) the simulator room became too warm, conditions being changed at the arrow. Later in the cruise phase (B), hunger was mixed with concern about airport location

Cockpit Monitoring and Alerting Systems

input to the thermoregulatory system was also measured (thermistor or thermography). Another approach would be precise control of the cockpit environment, but this is less satisfactory as operator control is usurped. Other sources of random noise (Figure 9.2 (B)) may not be as easily managed.

9.6 Characteristics of a future system

There are three parts to a future alerting system that are technology based and many parts that are not. The technology based components are human-centred and consist of the vigilance level detection system, the Pilot's Associate, and the vigilance assistance system. The non-technology based component consists of a management program which focuses on organisational values with respect to monitoring, as well as education and training in automation issues, autonomy, and optimisation of the cognate state. The non-technology based component is considered later (10.4-10.5). Each part of a future system will be described before the advantages, disadvantages and costs are briefly discussed.

9.6.1 System features

The three new technology components do not exist at present. The following is intended as a starting point.

Vigilance level detection A variety of sensors will measure a number of physiological parameters without electrodes, wires or harnesses. Some will be in close proximity to the pilot and some will be part of cockpit hardware and software. Those in close proximity will be in a glove, a wristband and the headset. Information transfer will be by telemetry. There must be no compromise with respect to the need for multiple measures and the absence of restraint. Much research and development in the area of non-invasive measurement in man is required.

Vigilance level processing The systems for producing a vigilance index are as little developed as the sensors. The skills base for developing suites of adaptive algorithms is available as is the computer processing power required for on-line index production and manipulation. The vigilance index will be produced in the Pilot's Associate. System properties similar to those being evolved for the Pilot-Vehicle Interface (PVI) sub-system of the Pilot's Associate will be needed (Judge, 1991; Geddes and Hammer, 1991), and many adaptive automation techniques will be used (Morrison et al., 1991; Emerson and Reising, 1991). The research required to link and manipulate vigilance measures has started in a very preliminary way (Fakhar et al., 1992).

Vigilance assisting system The modified Pilot's Associate system will optimise monitoring. With use, the Pilot's Associate will determine the optimal vigilance measures of each pilot for each task, being able to cater for general factors such as circadian rhythm shifts, stage in the duty cycle and prior workloads. The vigilance index will evolve interactively using statistical process control techniques, but will always retain a flexible component driven by performance and pilot input.

This approach will determine areas of monitoring strength and areas of monitoring weakness. If activated, it will determine the vigilance status of the pilot by comparing immediate measurements to the pilot's prior measurements for that particular flight phase. It will also be able to determine the amount of monitoring reserve. The Pilot's Associate will display monitoring status, monitoring reserve and possible strategies, allowing the crew to plan cockpit activities contingent on the available monitoring resources.

9.6.2 Advantages

There are a number of obvious advantages. For cockpit crew a key advantage is the opportunity to obtain job satisfaction while monitoring in automated cockpits. Job satisfaction will occur because cockpit crew will know about their monitoring and will have the tools to optimise it. In sharing control of the monitoring task, they will regain a measure of autonomy and develop new relationships with automated systems. An aspect of the automated systems used in the cruise phase of flight will become human rather than task-centred.

Even more important for cockpit crew will be the delineation of roles and the consequent reduction in role ambiguity. In addition, cockpit crew will be able to manage vigilance stress because of the presence of resources to optimise the process of vigilance. These two steps will be significant in reversing the effects of peripheralisation. The optimising of the human resource in the monitoring role has a place in minimising decreases in situational awareness.

There are multiple advantages for other stakeholders. The monitoring function will be optimised with safety implications for everyone. Airline operators will know that cockpit crew costs are justified because of defined rather than presumed activities. The possibility exists that for some segments of each flight the cockpit will contain a single crew member providing the Pilot's Associate is being used. This will promote much better rest for the non-flying pilot, a key issue in extended ultra-long haul flight.

9.6.3 Disadvantages

There are potential disadvantages of the future alerting system related to health, occupational health and new error forms.

Health The approach proposed will allow those who are interested and are curious about themselves, and unfortunately those who are preoccupied by their health, to track a variety of indices related to key body systems with unprecedented accuracy. Some with a predisposition to hypochondriasis and abnormal illness behaviour may struggle. These unfortunate individuals must be very rare in flight operations, considering existing selection systems.

Occupational health It is inevitable that the use of the alerting system will detect early evidence of a variety of disorders related to the cardiovascular and nervous systems. Some will posit that these changes are work related, because they have first been made visible in the workplace. While this is not a major problem, it is best managed proactively.

At issue is whether this information should be imparted to the pilot and or the organisation. It is essential that if these systems are to fulfil their role and be human-centred, then there must be no outside utilisation, nor presentation of health status. Thus, there must be no storage of data on any medium that might leave the aircraft outside the crew's control and cockpit memory devices must erase on power down. As the performance of the alerting system is partly dependent upon developing a personal file, a 'vigilance signature' as it were, it will be essential that each crew member takes responsibility for their information in a downloadable form at the end of each flight.

The Ideal Alerting System

New error forms Automation in cockpits has been associated with new error forms. It is likely that the alerting system will also produce new error forms, although they are less likely to be due to peripheralisation because of the human-centred nature of the system.

9.6.4 Costs

There are significant cost issues related to the fitting of alerting systems to road vehicles. A competent system could easily cost as much as the rest of a car. The cost of consequences, factored by the probability of a vigilance related accident and the probability of the vigilance system having preventing such an accident makes incorporation of vigilance systems in private road vehicles a dubious exercise. The cost of consequences is likely to be too low to justify alerting system technology, even if the full societal costs of accidents are included.

Cost issues are dramatically affected by machine size in each transport system. Thus, while the cost of a vigilance assisting system remains relatively constant and is only minimally affected by crew size, the cost of the consequence from a vigilance related incident or accident increases in a non-proportional manner. Thus, the costs of a vigilance related accident involving a future 'mega-carrier' during descent in a populated area are so large that the costs incurred in fitting an alerting system are insignificant.

9.7 Summary

There are a number of deficiencies in present alerting systems. In present systems the control of the alerting system does not reside with those required to sustain attention. This reflects the design philosophy common to present systems where the emphasis is on performance at the expense of the person. Not surprisingly, such systems are either not used, or are bypassed by human ingenuity, nullifying their whole purpose. They incur costs without the cost reduction consequent upon reducing errors. Another deficiency of present systems is that they recurrently remind operators about the vigilance issue without giving them the means to measure vigilance or do anything to alter their vigilance level. It is difficult to think of a better way of producing vigilance stress.

Cockpit Monitoring and Alerting Systems

A damming defect in present alerting systems is that they interpose unrelated functions between the monitor and the monitoring task. This aspect of present designs is belittling to those in the monitoring role, stimulating part of the workforce to develop work practices that bypass the alerting system entirely. The failure to produce a synergy between the wishes of designers and operators is remarkable, particularly when the safety of those in the monitoring role is often involved.

The ideal alerting system should focus on measuring vigilance level without impediment, sampling sufficient parameters so that the activity in relevant neural networks can be assessed. These outputs will be converted by the Pilot's Associate into an index, which the pilot and the Pilot's Associate will create and use interactively in managing monitoring.

Part C
Monitoring Management

10 Monitoring management, interim and future changes

The presence of the person performance system in the Boeing 747-400 (8.4) suggests that many are perturbed by the monitoring problem in cockpits. The Boeing system is likely to prove unsatisfactory (8.5.2). While a human-centred technology is developed, an interim monitoring management plan is required. The aim of the plan is to reduce peripheralisation effects on monitoring by decreasing vigilance stress and role ambiguity (6.1-6.5). After the effect of flight phase is discussed, interim management changes are considered.

There are more fundamental long term solutions to the vigilance problems produced by peripheralisation effects (6.5). Solutions should involve the incorporation of a new human-centred technology (9.6), continuation of an interim change program in both the organisation (10.4) and the cockpit (10.5), and reexamination of the human role in automation systems (10.6.1). Although the cruise phase will be mainly considered, monitoring in other flight phases will also be addressed.

10.1 Flight phase

There are a number of different phases in any flight and the monitoring task differs in each phase. In commercial passenger flying there are preflight, taxi, take-off, climb, cruise, descent, approach, landing and shut-down phases. Each has different monitoring requirements. A change program should focus on areas of immediate need.

10.1.1 Accident incidence and implications

Flight incidents and accidents occur predominantly during descent and landing, the cruise phase being one of the safest in commercial aircraft operations (Nagel, 1988). The difference in accident occurrence is very dramatic, the descent and landing phases accounting for 59 per cent of all accidents while the cruise phase only accounts for 9 per cent (Nagel, 1988). However, it is a mistake to equate flight phase with accident cause as there are many situations where the activity in one phase has profound effects on subsequent phases.

Even the preflight phase can effect later stages, as evidenced by the fuel starvation incident involving the Air Canada Boeing 767 (2.6.3) and others. In the Mount Erebus disaster, Air New Zealand's navigation section gave the crew a mismatched set of navigational coordinates before they were even in the cockpit (O'Hare and Roscoe, 1990). Thus, the flight phase in which an accident occurred has not been a good marker of the phase in which a problem commenced.

10.1.2 Cruise phase

The cruise phase dominates the flying time of the crew in widebody aircraft. Monitoring is the dominant and often the only cockpit activity. There is no proof that crews, who are carrying out monitoring during the cruise phase, suffer impaired performance in subsequent flight phases. However, there is a lot of circumstantial evidence which suggests that a proportion of crews are leaving the cruise phase in a relatively dysfunctional condition compared with when they entered it. All the error promoting factors can be implicated (2.5.1-2.5.2) in cruise phase performance degradation, but monitoring, while being in a peripheralised state in an automated cockpit, is probably the key culprit.

It is proposed that monitoring during cruise in modern cockpits has effects on crew members and these effects last throughout the phases of cruise, descent and landing. Support for this proposition includes:

1   Human factors accidents in 'faultless' aircraft (3.4.2) suggest that many descent phase accidents were primed long before the descent phase started. There are many incidents to support this.

2   Vigilance level changes significantly within half an hour of the start of the cruise phase (7.2.3). These changes occur pari-passu with changes in communication (Cabon et al., 1991), suggesting that degradation in vigilance is intimately linked to reduced performance in other activities.

Thus, the cruise phase, outwardly a period of relative safety, may actually be the phase where peripheralisation effects are wreaking their greatest havoc. Peripheralisation effects are insidious and the behavioural changes induced by the automated cocoon probably start early, increasing gradually throughout the rest of the flight. Cruise phase monitoring is the area of greatest concern for monitoring management programs.

10.1.3 Cruise phase focus

Reversing the problems associated with prolonged monitoring is a worthy activity in its own right, particularly as cockpit crew will spend more and more time monitoring as automation progresses. It has been suggested (10.1.2) that an added reason for changing monitoring is because of its intimate links with peripheralisation effects, vigilance stress and role ambiguity (6.5). Monitoring management in the cruise phase will be considered in detail because small changes achieved in this phase have the potential to significantly affect subsequent phases. Monitoring in other phases is considered later (10.7).

10.2 Breaking the loops

The peripheralisation-vigilance cycle (6.1) is best slowed by reducing vigilance stress and optimising vigilance performance (6.5). In the longer term, significant improvement in vigilance performance will require new forms of alerting system (Chapter 9). As an interim measure, a monitoring management plan can reduce vigilance stress, although its effects are likely to be small.

The peripheralisation-stress cycle (6.3) can be altered by a number of means. An organisational ethos, which focuses on the key role of monitoring and resources it appropriately, will provide the right milieu for tackling the problems of role ambiguity (6.2.1) and cognitive stress in the cockpit. An organisational commitment to transfer control of monitoring processes to the crew is also essential. Role ambiguity will be reduced in time by technological developments (Chapter 9) which are human-centred rather than task-centred. Reductions in role ambiguity will slow the peripheralisation-stress cycle.

10.3 Present cruise phase management

There are a number of ways that cruise phase monitoring is managed at present. These include crew training, crew rotation, CRM programs and LOFT sessions.

10.3.1 Crew training

A great deal has been written about aircrew training. The arguments related to the amount of training required with respect to increased automation, and the specific argument that automation reduces the need for traditional flying skills is not directly relevant to the issue of training for the monitoring role. There is little doubt that the training in standard operating procedures (SOPs) encompasses both in practice and in spirit the needs for systemised monitoring of aircraft function. However, the 'monitoring approach' (Orlady, 1989) must be implemented properly, the 'theory espoused' must also be the 'theory practiced'. This matching of intent and practice must permeate all levels of the training hierarchy if the training process is to be successful.

Despite the guidelines set by SOPs, surveys of pilots' opinions suggest that monitoring is bedevilled by not knowing the 'design intent' of systems. A common complaint is that it is not possible for the monitor to understand how the system being monitored is set up to accomplish its specific task. Pilots identify shortfalls in the amount of systems information available during training (Orlady, 1989). Common pilot problems in advanced cockpits relate to why a cockpit system behaves in a particular manner, what is its present behaviour and what will happen next (Orlady, 1989). Insufficient and inappropriate information has peripheralising effects, particularly on situational awareness (2.3.3). It is unknown how manageable this is by training.

The inability of training to alter the management of automated systems is easily illustrated. Pilots flying the Boeing 757 can be faced with the problem of wanting to descend before the programmed descent point. Pilots bypass the aircraft computer's flight plan by entering a fictitiously high tailwind to induce the automated system to start the descent early (Wiener, 1989). The input of misleading information suggests that training in automated system management is partly ineffective.

10.3.2 Crew rotation

Some airlines rotate crew after a set interval (90 minutes) on long sectors, spare crew relaxing in the adjacent cockpit rest area. The advantage of this approach is that it tends to spread the monitoring task evenly. The disadvantage is that it is a disruptive procedure, cockpits not being designed for crew rotation. Crew rotation ignores individual monitoring abilities and ensures that all crew have their ability to sustain attention blunted in some respect.

Unexpected but useful effects of crew rotation are the arousal changes that occur as two to three medium sized humans try to swap seats in the limited space of a cockpit. The physical activity and the care required during cockpit movement alter arousal levels transiently.

In ultra-long haul operations, some airlines use two separate crews divided into 'A' and 'B' teams, the former carrying out take-off, climb, descent, approach and landing while the latter carry out all the cruise phase activities. This approach minimises costs with respect to crewing levels, but may fail to produce effective monitoring in the cruise phase. It is difficult to imagine a more unfulfilling task than that faced by 'B' team members. Either they do very little monitoring, or they must suffer role ambiguity and vigilance stress.

10.3.3 Crew numbers

At present, single person operation is not allowed except under exceptional circumstances. In the two crew aircraft this restriction complicates cockpit management, particularly when crew are required in the cabin and when there are additional company tasks. The issue of crew numbers and the arguments for two versus three man crews is complex and far from settled from the pilots' point of view (Wiener, 1989).

10.3.4 CRM and LOFT programs

Monitoring is not a specific feature of CRM programs because monitoring is such a primary function that it is usually handled at the SOP level. The scepticism that many operators have about some aspects of CRM probably inhibits them from committing the monitoring function to programs which are not entirely of their own making

Cockpit Monitoring and Alerting Systems

(Lautman and Gallimore, 1989). Crew coordination and decision making activity is practiced in LOFT sessions. Monitoring issues may be considered by crews as they debrief themselves.

10.4 Interim changes in the organisation

Peripheralisation effects may be attenuated by bringing about a number of changes in organisational values and practices.

10.4.1 Organisational values

The monitoring role of cockpit crew is not a function which airlines have been concerned about. Few organisations have determined what monitoring resources they have and what they really require. Monitoring management has been minimal, usually consisting of training in effective instrument scanning as part of a SOP. While airlines are probably less deficient in their commitment to monitoring management than many other organisations which have large monitoring demands, the shift towards longer sectors, larger aircraft, and more automated cockpits suggests that organisational priorities will have to change. There must be a shift in priorities towards resourcing and facilitating monitoring activities and this must be endorsed by all levels of airline management.

At present the control of the monitoring function is not clearly defined. While the cockpit crew are responsible for monitoring, the rules which govern how monitoring is carried out, when it is carried out, and what is monitored are set by the airlines' flight management teams. The breaking of the peripheralisation-stress link requires that those monitoring have freedom to choose the style of their monitoring, the amount of ancillary aid they want to use and how they pool their resources. Flight operations must come to terms with relinquishing control of monitoring management.

10.4.2 Crew training and education re monitoring and automation

Flight operations need to determine the prevailing attitudes, knowledge and skills with respect to monitoring, automated systems and information management. Training programs must be reviewed with respect to these entities.

Surveys Information is essential for altering attitudes and can be obtained by surveys and in LOFT programs. The tracking of attitudes to automation and peripheralisation allows change processes to be monitored. Thus, 200 Boeing 757 pilots agreement or disagreement on a five point intensity scale to the statement, 'I look forward to more automation - the more the better' was symmetrical around a neutral response in two surveys one year apart (Wiener, 1989). Other surveys have produced different responses, depending upon training and experience (James et al., 1991). Surveys are essential for detecting shifts in attitudes concerning monitoring, peripheralisation and automation.

Information and monitoring management The expectation that aircrew can absorb more and more information during training is unrealistic and inappropriate considering the move towards skills-based programs. While the electronic library will be a feature of future aircraft flight decks, the ability to source information efficiently cannot always be left to software systems. Aircrew training must optimise the individual's strategies for efficient information retrieval as well as for the general management of information. Both are required if those in the monitoring role are to achieve job satisfaction. The attitudes of crew to automation and monitoring will change as crew autonomy and adequate resourcing of monitoring activities become part of flight operations policy.

10.5 Interim changes in the cockpit

There are a number of specific changes in cockpit activities that will immediately assist those faced with the monitoring task.

10.5.1 Autonomy

Pilots have generally positive perceptions about flight deck automation. A case can be made for giving flight deck crew the option of using the resources as they choose. Thus, even in the 'poorly' equipped cockpits of today, flight crew have choice although SOP do not allow them to exercise it.

Autonomy At present the control of the monitoring function is not clearly defined. As discussed previously (10.4.1), flight operations determine the criteria for monitoring, but the issue of control and stress reduction require that there must be changes. Crew must be able to choose monitoring styles which they find effective and which they are comfortable with.

Turning automated systems off It has been recommended that crews should be allowed, possibly trained, to 'click it off' and make partial use of automated systems. At present there are no guidelines as to how this freedom should be used in cruise, given that the aircraft has to be 'tuned' to optimise fuel burn and the economics of the whole flight. The recent Boeing 747-400 modification by Cathay (altering engine thrust angles by one degree for ultra-long haul flights) suggests that the variables that can be manipulated during any human intervention in cruise may damage the economics of the operation. Nevertheless, the opportunity to use automation flexibly is important in giving flight crew control, reducing monitoring stress and promoting vigilance.

10.5.2 Crew selection and team formation

A significant issue in monitoring management is the requirement to form new crews with each duty cycle. Monitoring quality and capacity would be increased if stable teams were part of normal commercial operations. Stable teams have many advantages (2.5.1).

Present crew selection At present crews select their flights by a bidding system where seniority is the primary ranking factor. One consequence in a large airline is that it is relatively unusual for a crew to have ever been together before. There are benefits and disadvantages to this system, but the benefits relate more to personal freedoms and hierarchical privilege than optimising human resources and producing job fulfilment.

The disadvantages of forming new teams for each duty cycle are clear and have been demonstrated by cockpit research studies (Chidester and Foushee, 1989) which show that crew coordination and function improve through a tour of duty. Military aircrews always work as teams, having trained as teams and carried out mission simulations, which are the equivalent of LOFT exercises, as teams. Crew coordination is an important factor in improving cockpit monitoring.

Teams, monitoring quality and capacity The formation of teams by commercial aircrews would increase the monitoring quality and capacity of each crew. Monitoring quality would be improved because each person would have some idea of the expected state of alertness and monitoring performance of the other. This is no more than what the Pilot's Associate will be doing when its development is complete. The continuous assessment of crew member attentiveness and performance is a practical possibility and training could occur within CRM programs. People who have a significant amount of shared work experience develop a template of their companions' work performance, which is an aggregate of past satisfactory performances. This assessment of alertness and performance would be very sensitive to deviations, the early detection of alertness decline, and the state of internal well-being which is part of role competence.

Monitoring capacity would be increased because the crew could obtain the maximum amount of useful monitoring out of each individual. Thus, instead of crew rotations occurring by the clock, crew rotations would be driven by the supply of monitoring resources.

Utilisation of human resources There are other benefits of stable crews. One of the roles of LOFT exercises is to bring together crews to practice flying particular sectors in a simulator. Crews carry out their own critique of the video replay of their performance. LOFT exercises are an important adjunct to training and a critical extension of CRM programs, but they are expensive in man hours and simulator time. In part, the issue being handled by the LOFT exercise is the melding of individuals into functional teams, a part of the melding being non-procedural and interpersonal. The need for this training and consumption of pilot hours would decline if cockpit personnel formed teams as is done in military aviation. A decline in this use of pilot hours would allow airlines to employ less pilots.

Resistance Resistance to change will arise with altering crew selection from the seniority-based scheme. Resistance can be partly nullified by combining some aspects of the constant crew and the seniority systems. Thus, seniority ranking in the company might not be changed by flying as a constant crew, but flying as teams could either give advances in seniority purely for the sake of duty choice, or give teams the opportunity to fly specific aircraft. Those willing to fly as teams could be channelled to aircraft where team coordination was a more critical factor. These aircraft are more automated, have a greater tendency to peripheralise their crew, and all have a two crew cockpit. The present method of crew selection is more vulnerable with two crew cockpits, because there is the potential for 'impasse' situations related to personality and procedural preference. This possibility is to be avoided in long haul operations if at all possible. Stable crews would virtually eliminate it.

10.5.3 Napping

Although not officially sanctioned, napping is part of cruise phase flying. Almost all naps are unplanned and are probably underestimated in their frequency (Dinges and Graeber, 1989; Beech, 1991). Napping has been found to be an issue for ship crews, train drivers, telephonists and car drivers. It has been suggested that napping does not occur to all cockpit crew members simultaneously (Beech, 1991). However, there are incidents and anecdotal reports of whole crews being asleep for extended periods of time.

Some have proposed that there are significant benefits from programmed napping (Dinges and Graeber, 1989). Napping is likely to affect monitoring ability significantly, although this has not been proved. The NASA group have used instrumented crew members in a commercial widebody flight, and demonstrated beneficial effects on fatigue levels, and presumably on performance, when a crew member was given time to nap, the optimum nap time being 40 minutes with a 5 minute nap preparation time and a 20 minute post nap recovery phase. This programmed napping was considered to be beneficial up to half an hour before a descent phase occurred.

It has been observed that some crew members are frequent and easy nappers while others never nap. Random crew selection will complicate the use of napping as a crew of two reluctant nappers will not be able to use this technique optimally. Stable crews could accommodate these differences.

10.5.4 Stress management

The present increased interest in stress management by CRM program managers needs re-examination. While there is no harm in teaching

Monitoring Management, Interim and Future Changes

aircrew about life stress management, ineffective as it may be, it is a mistake by aircrew and aircrew training organisations to feel that these type of programs can contribute anything to the management of the stress caused by role ambiguity or vigilance tasks. There is a danger that training organisations will use their stress management programs as a simple fix. In time this will discredit many of the real benefits of CRM.

Effective management of role ambiguity lies in empowerment and shifting the focus of control with respect to monitoring from outside decision makers to those who are directly involved.

10.5.5 Optimising the cognate state

Monitoring requires knowledge about the systems being monitored. If the ability of cockpit crew to absorb information is suboptimal, the provision of more information via an electronic library is not a realistic solution to management of information in automated systems. The ability of crews to monitor is susceptible to a multitude of factors which impair information absorption and processing.

Each crew member is well aware of the need to optimise their physical and mental state. The cognate state of the group is more fragile than that of the individual, and all crew members must think of the overall ability of the crew to sustain attention if monitoring resources are to be optimised. Awareness of fatigue, boredom, change in communication style and content, difficult monitoring conditions (flight phase), adverse physical factors (temperature, noise) should alert crew to difficult monitoring conditions. Crew members will understand enough about monitoring to continuously optimise their own monitoring ability and that of their partner.

It is very likely that the ability to monitor efficiently varies widely, and the present method of selection of flight crew has no bias in this regard. Most human abilities differ by very significant amounts, normal ranges often having an upper value which is more than twice the lower value. Sometimes the difference is very much greater. It is likely that cockpit crew have significantly better monitoring skills than the rest of the population, but still have very significant differences in monitoring ability amongst themselves. At present there is no means of measuring this ability usefully, but it is likely that cockpit crew have some inkling of whether they are relatively good or bad at monitoring. Crews should be encouraged to optimise their combined monitoring resources.

10.6 Future changes

Future changes involve the incorporation of a human-centred alerting system in the cockpit (9.6). Successful implementation will partly depend on the change in attitudes and skills produced by the interim programs (10.4-10.5). At the same time there must be a review of the processes underlying the automation of control systems in which human monitoring is required. The key elements in such a review are considered before the impact of the technological and non-technological aspects of alerting systems are discussed.

10.6.1 Refashioning views on automation and monitoring

Guidelines for the automation of control tasks and the fashioning of the monitoring role have been proposed (Wiener and Curry, 1980; Wiener, 1989). Features which are essential for the automation of systems in which human monitoring is required must include:

1	The operation of systems being monitored must be transparent and not opaque to the operator. System degradation must be graceful and involve the operator.

2	The monitoring system must be malleable and sensitive to the needs of the operator, the state of the operator, as well as temporal, cultural, sleep cycle and fatigue factors.

3	Those in critical monitoring situations should have their alertness level and their monitoring performance measured, and these measures should be available to the operator for interactive modification of the monitoring role.

4	The monitor must be provided with the means to alter monitoring performance.

5	Monitoring is as important a task as any other in automated systems, and motivation, training and evaluation is essential.

5	There is no place for additional duties when monitoring demands are low. The interaction between those monitoring and the

Monitoring Management, Interim and Future Changes

monitoring system should be directed at optimising vigilance levels and harbouring the monitoring resource.

6 Assessment of the monitor's state of alertness should be continuous, non-invasive and never involve a secondary task.

Implementation of these recommendations will reduce role ambiguity as a cause of monitoring stress and break the peripheralisation-stress loop. Better monitoring and less vigilance related errors will slow the cycling of the peripheralisation-vigilance loop. Better monitoring should become increasingly cost effective (9.6.4), particularly in larger aircraft.

10.6.2 Consequences for cockpit operations

Various characteristics of the new attention indicating system have been discussed (9.6.1-9.6.4). A vignette of how the system might affect a two man crew in the first phases of flight is presented.

Pre-cruise phases Both crew members will put their personal data discs into the PVI sub-system of the Pilot's Associate (9.6.1) and will put on their head sets. After take off and towards the end of ascent, one of the crew, who has elected to use the alerting system, will put on a glove or wristband and input his wishes with respect to monitoring style and assistance level. He may be prompted to readjust the position of his microphone boom, wristband or glove. After determination of his monitoring reserve, selection of vigilance level display, and discussion of prospective monitoring issues, his companion leaves the cockpit for the rest area.

Cruise phase The pilot flying (PF) has specialised in monitoring management and has modified the algorithms determining his vigilance level to use only two sensor measures, both sensors being contained in the microphone boom. He discards the wrist band after checking with the PVI sub-system that there is nothing unexpected in his vigilance status. He continues his instrument scan, which includes a vigilance level and vigilance performance display. After some hours the PVI sub-system draws his attention to a greater than expected vigilance decrement at half flight distance. He contacts the non-monitoring pilot and they elect to use 'napping' as a means of expanding their monitoring resource. The monitoring specialist naps in the cockpit, the PVI sub-system measuring the efficacy of the nap while the other crew member takes over the PF role. She elects not to use the alerting system. The new PF is not a napper and she knows she is less suited to the monitoring role. Her prowess lies in the management of the take-off, ascent, descent and landing phases. The PVI sub-system monitors the new PF by her performance, posture and respiration without any person-machine interaction. She wakes the napping pilot when the PVI subsystem provides a prompt. Once vigilance levels become satisfactory post-nap, she discusses the next phase, the division of duties, and then returns to the rest area.

Cockpit crew will need to understand vigilance networks and the processes underlying the options available to them when interacting with the alerting system. Much of the training will be carried out using interactive skills-based packages running on sensor equipped terminals. Their personal discs will be used from the start of their training and simulators will not be required.

10.6.3 Consequences for the organisation and others

Flight operations management will be intimately involved in implementation of the alerting system, management personnel being required to be competent in its use. Education of management will be critical. The present push by airlines for the 'paperless' cockpit, without concern for the cognate state of crews, suggests that management may be insensitive to human capabilities (Bailey, 1991).
  The involvement of human factors specialists in aviation suggests that aircraft design groups and certification authorities are aware of their ability to produce improvements in efficiency and prevent human error. However, it is not clear that the aircraft design groups and the certification authorities have developed a satisfactory philosophy concerning flight deck automation (Norman et al., 1988; Wiener, 1988). It comes as some surprise that the recent developments by Boeing (8.4) and those planned for the ATF (8.6.2) are only performance measurement systems and are basically flawed (8.5.2). Human factors specialists involved in cockpit design and certification will be affected by the change in values inherent in the new alerting system.

10.7 Alerting system in other flight phases

Some phases of flight have higher workloads. The new alerting system is able to measure the amount of effort (7.7) being spent on a particular task and compare that with the individual's prior attempts. Effort, or arousal task mismatch can probably be detected.

In many situations, this information is not particularly useful as crew changes are either inappropriate or impossible. Thus, an unexpected event, such as windshear, engine fire or a runway change in a complex air traffic control environment, produces high demands on a crew. The new alerting system may indicate an inappropriate level of arousal and effort, an indicator of impending poor performance. Providing knowledge of the mismatch without the means of doing something about it is counterproductive.

It is unlikely that the management of the unexpected event is going to be altered by the new alerting system. It has been suggested that these situations are best handled by simulator practice (Green, 1985), although it is accepted that competence or incompetence on the simulator does not always translate into identical performance in an aircraft. However, the alerting system will make some difference to the management of the unexpected event because the crew should be able to harbour their monitoring resources and other coping abilities. Thus, an undoubted role for the alerting system is ensuring that the ability of individuals is optimised before entering a high workload flight phase.

10.8 Alerting systems in other monitoring roles

The costs to companies of monitoring failure is at present not incorporated into management decision making because most non-airline monitoring failures only harm humans indirectly. A large tanker spill, due to faulty monitoring on the ship's bridge, is devastating to the environment but lacks the voyeuristic immediacy of pilot error. A failure to pick a trend on a financial exchange display is expensive but can be covered by customer charges. An error in the monitoring of a power grid can be passed off as an environmental or demand prediction problem. It appears as if the passenger aircraft cockpit is the work place where the monitoring role has been of greatest concern.

The indifference to the importance of monitoring outside aviation is unlikely to remain. It can be predicted that a variety of monitoring failures outside the airline industry will incur increasing societal costs due to consumer awareness. It is on this background that the monitoring function will become more critical. The alerting system and its underlying design principles can be employed in a variety of the more critical monitoring roles. Indeed, it is likely that an alerting system would make organisations better off in terms of human resource management and their cost structure.

While the alerting system has been discussed in the context where there is a limit on the amount of human substitution that can occur, the presence of a larger pool of people (in a nuclear power station) does not necessarily provide a cost effective alternative. The essence of the new alerting system is maximising the monitoring ability of each individual. Substitution is only an effective solution providing it is carried out when the existing human resource has been fully used.

10.9 Summary

It is not possible to immediately alter all the circumstances of those in critical monitoring roles. New technology which humanises and enhances monitoring processes will take time to develop. There is need for immediate changes, even if they only provide partial solutions. Interim solutions for the monitoring problem in the cruise phase of flight should include:

1. Shifting control of the monitoring role to the cockpit.

2. Altering the priority given to monitoring within the organisation.

3. Training in automated systems management, information management and monitoring processes and the tracking of attitudes and skills with recurrent surveys.

4. Examination of the methods of crew formation with a view to promoting stable crew structures in cockpits where peripheralisation is a problem.

5. Encouragement of napping and other autonomous activities.

Monitoring Management, Interim and Future Changes

6   Avoidance of stress management programs as solutions to cognitive stresses in cockpits.

7   Care with the use of existing cockpit person performance systems.

The above program is essential as a precursor to the introduction of the alerting system, which unheralded might meet with unreasoned resistance. At the same time as the content and skills of the above programs are being assimilated, the necessary research and development of the sensors and the Pilot's Associate can occur.

11 Conclusions

There can be little doubt that replacing human functions with machine functions is necessary, useful and beneficial. Replacing a human function while asking a human to watch a machine performing that function is also going to be necessary, useful, but not as clearly beneficial to all. At present, the role of machine watcher is ill defined. The modern commercial airline cockpit is the workplace in which the monitoring of automated systems is being refined and the guidelines for a large array of monitoring roles are being brought into focus. At present, it is unknown if the human will become a token, bypassed, helpless and sometimes hapless. It is possible that humans will give up their involvement with control systems and accept the consequences of the small error rates that are part of fully automated systems.

It is also possible that humans will not just be tokens, but human inadequacies in sustaining attention complicate the role of machine watcher. While this role remains undefined, there is a problem for those tasked with being vigilant on others' behalf. Is it possible that the monitoring problem has been overstated? It is possible, but it is unlikely. The International Civil Aviation Organisation, in commenting on the explanations behind the shooting down of a Korean Airlines Boeing 747, said 'Each of the postulations assumes a considerable degree of lack of alertness on the part of the entire crew, but not to a degree that is unknown in international civil aviation' (Wiener, 1988).

This book attempts to outline an approach to monitoring management. The approach is pragmatic, as theory alone has helped little in this area in the past. In the absence of appropriate technology, this book suggests that the focus in monitoring management should be shifted from machine measurement of monitoring performance back to the humans monitoring, aiming to remove some of the obvious anomalies. The major anomaly is the predilection of planned monitoring assistance systems to measure performance alone, without any awareness of the state of the person monitoring. This quality assurance approach is a modern version of 'time and motion' assessment, is outdated, and will be treated with disdain. The ingenuity of man to work around and bypass systems which control and dictate work activity, but ignore the person, will make the systems under development a costly mistake. Another mistake in present cockpit monitoring is the inefficient use of human resources. The easiest way of improving the present system is to shift crew selection from the present individual bidding system to a team selection method. Napping, flight deck autonomy and reduced control by flight deck management of the monitoring function will also help.

In the longer term monitoring management requires new technology, but this technology must be focused on humans, be controlled by humans and be designed to restore the balance in the relationship between man and machine. It is not acceptable for pilots to feel that in their workrole they are 'along for the ride'. This lack of role definition for those monitoring will be resolved when their own level of vigilance becomes their unique contribution. Outwardly it might appear that the solution to the monitoring problem involves yet another layer of technology being interposed between the human and the machine. This new technology is acceptable because the sensors and the processing system are human rather task-centred. This technology may well assist in unravelling part of the task-centred component of existing equipment.

This book can only give some details about the new alerting systems and cannot completely describe the technology nor the future modes of use. However, the guidelines and principles for such a system are clear. Much research and development remains to be done, there being little doubt that sensors can be made which will remove the vigilance researchers' objections of clumsiness, inflexibility and unreliability. Nor is there doubt that there is already sufficient machine intelligence to restore a role for humans in monitoring. Will the new alerting systems be used, or be bypassed as current alerting systems have been? It is impossible to be sure and it is likely that some will try to 'trick' the alerting system. However, most of the measures which will be used to produce a vigilance level are not under conscious control and synchronous manipulation of multiple parts of the autonomic nervous system is not possible. More importantly, it is likely that the alerting system will be seen as a useful adjunct in the cockpit, like a moving map display, where the 'design intent' and the wish of those monitoring are at one. Such synergy can only be beneficial for those in the monitoring role and all of those affected by their monitoring efforts.

References

Anderson, K.T. (1990), 'Arousal and the inverted-U hypothesis: a critique of Neiss's "Reconceptualising arousal"', *Psychological Bulletin* 107, pp. 96-100.

Alkov, R.A., Gaynor, J.A. and Borowsky, M.S. (1985), 'Pilot error as a symptom of inadequate stress coping', *Aviation, Space and Environmental Medicine* 56, pp. 244-247.

Averill, J.R. (1969), 'Autonomic response patterns during sadness and mirth', *Psychophysiology* 5, pp. 399-414.

Baker, C.H. (1959), 'Attention to visual displays during a vigilance task: II. Maintaining the level of vigilance', *British Journal of Psychology* 50, pp. 30-36.

Bailey, J. (1991), 'Towards the paperless cockpit', *Flight International* No.4277, 140, pp. 30-31.

Balick, L.R. and Herd, J.A. (1986), 'Assessment of physiological indices related to cardiovascular disease as influenced by job stress', *Journal of Organizational Behavior Management* 8, pp. 103-115.

Ballas, J.A., Heitmeyer, C.L. and Perez, M.A. (1991), 'Interface styles for adaptive automation', in Jensen, R.S. (ed.) *Proceedings of the Sixth International Symposium on Aviation Psychology*, pp. 108-113.

Beatty, J. (1982), 'Phasic not tonic pupillary responses vary with auditory vigilance performance', *Psychophysiology* 19, pp. 167-172.

Beech, E. (1991), 'Caught napping', *Flight International* No.4278, 140, pp. 30-31.

Bedeian, A.G. and Armenakis, A.A. (1981), 'A path analysis study of the consequences of role conflict and role ambiguity', *Academy of Management Journal* 24, pp. 417-424.

Beehr, T.A. and Franz, T.M. (1986) 'The current debate about the meaning of job stress', *Journal of Organizational Behavior Management* 8, pp. 5-18.

Beh, H.C. (1990), 'Achievement motivation, performance and cardiovascular activity', *International Journal of Psychophysiology* 10, pp. 39-45.

Beringer, D.B. (1989), 'Exploring situational awareness: a review and the effects of stress on rectilinear normalization', in Jensen, R.S. (ed.) *Proceedings of the Fifth International Symposium on Aviation Psychology*, pp. 646-651.

Berrichi, H., Tibergé, M. and Arbus, L. (1991), 'Détection de la vigilance: enregistrement simultané des signaux E.E.G. et E.O.G.', in Vallet, M. (ed.) *Le maintien de la Vigilance dans les Transports*, L'Institut National de Recherche sur les Transports et leur Sécurité, Caen, pp. 143-150.

Billings, C.E. (1989), 'Toward a human-centered aircraft automation philosophy', in Jensen, R.S. (ed.) *Proceedings of the Fifth International Symposium on Aviation Psychology*, pp. 1-7.

Billings, C.E. and Reynard, W.D. (1984), 'Human factors in aircraft incidents: results of a 7-year study', *Aviation, Space and Environmental Medicine* 55, pp. 960-965.

Blake, R.R. and Mouton, J.S. (1985), *The managerial grid* III, Gulf, Houston.

Bowers, C.A., Morgan, B. and Salas, E. (1991), 'The assessment of aircrew coordination demand for helicopter flight requirements', in Jensen, R.S. (ed.) *Proceedings of the Sixth International Symposium on Aviation Psychology*, pp. 308-313.

Brief, A.D., Burke, M.J., George, J.M., Robinson, B.S. and Webster, J. (1988), 'Should negative affectivity remain an unmeasured variable in the study of job stress?', *Journal of Applied Psychology* 73, pp. 193-198.

Broadbent, D.E. (1958), *Perception and communication*, Pergamon Press, New York.

Broadbent, D.E. (1978), 'The current state of noise research: reply to Poulton', *Psychological Bulletin* 85, pp. 1052-1067.

Brown, C.E., Boff, K.R. and Swierenga, S.J. (1991), 'Cockpit resource management: a social psychological perspective', in Jensen, R.S. (ed.) *Proceedings of the Sixth International Symposium on Aviation Psychology*, pp. 398-403.

Bruning, N.S. and Frew, D.R. (1987), 'Effects of exercise, relaxation, and management skills training on physiological stress indicators: a field experiment', *Journal of Applied Psychology* 72, pp. 515-521.

Cabon, P.H., Mollard, R., Coblentz, J.-P., Fouillot, C. and Molinier, G. (1991), 'Vigilance of aircrews during long-haul flights', in Jensen,

References

R.S. (ed.) *Proceedings of the Sixth International Symposium on Aviation Psychology,* pp. 799-804.

Cannon, W.B. (1914), 'The interrelations of emotions as suggested by recent physiological researches', *American Journal of Psychology* 25, pp. 256-282.

Chandra, D., Bussolari, S.R. and Hansman, R.J. (1989), 'A comparison of communication modes for delivery of air traffic control clearance amendments in transport category aircraft', in Jensen, R.S. (ed.) *Proceedings of the Fifth International Symposium on Aviation Psychology,* pp. 433-438.

Chaput, D., Petit, C., Planque, S. and Tarrière, C. (1991), 'Un système embarqué de détection de l'hypovigilance', in Vallet, M. (ed.) *Le maintien de la Vigilance dans les Transports,* L'Institut National de Recherche sur les Transports et leur Sécurité: Caen, pp. 105-112.

Chidester, T.R. and Foushee, H.C. (1989), 'Leader personality and crew effectiveness: a full-mission simulation experiment', in Jensen, R.S. (ed.) *Proceedings of the Fifth International Symposium on Aviation Psychology,* pp. 676-681.

Clothier, C.C. (1991), 'Behavioural interactions across various aircraft types: results of systematic observations of line operations and simulations', in Jensen, R.S. (ed.) *Proceedings of the Sixth International Symposium on Aviation Psychology,* pp. 332-337.

Conley, S., Cano, Y. and Bryant, D. (1991), 'Coordination strategies of crew management', in Jensen, R.S. (ed.) *Proceedings of the Sixth International Symposium on Aviation Psychology,* pp. 260-265.

Cooper, C.L. (1985), 'The stress of work: an overview', *Aviation, Space and Environmental Medicine* 56, pp. 627-632.

Costley, J., Johnson, D. and Lawson, D. (1989), 'A comparison of cockpit communication B737 - B757', in Jensen, R.S. (ed.) *Proceedings of the Fifth International Symposium on Aviation Psychology,* pp. 413-418.

Craig, A. (1976), 'Signal recognition and the probability-matching decision rule', *Perception & Psychophysics* 20, pp. 157-162.

Czeisler, C.A. and Jewett, M.E. (1990), 'Human circadian physiology: interaction of the behavioral rest-activity cycle with the output of the endogenous circadian pacemaker', in Thorpy, M.J. (ed.) *Handbook of sleep disorders,* Dekker, New York, pp. 117-136.

Davies, D.P. (1979), *Handling the big jets,* CAA, London.

Davies, D.R. and Krkovic, A. (1965), 'Skin-conductance, alpha-activity, and vigilance', *American Journal of Psychology* 78, pp. 304-306.

Davies, D.R. and Parasuraman, R. (1982), *The psychology of vigilance*, Academic Press, London.

DeBoer, S.F., Koopmans, S.J., Slangen, J.L. and Van der Gugten, J. (1989), 'Effect of fasting on plasma catecholamine, corticosterone and glucose concentrations under basal and stress conditions in individual rats', *Physiology and Behavior* 45, pp. 989-994.

Degani, A. and Wiener, E.L. (1991), 'Philosophy, policies and procedures: the three p's of flight-deck operations', in Jensen, R.S. (ed.) *Proceedings of the Sixth International Symposium on Aviation Psychology*, pp. 184-191.

Denoyer, M., Sallanon, M., Buda, C., Kitahama, K. and Jouvet, M. (1991), 'Neurotoxic lesion of the mesencephalic reticular formation and/or the posterior hypothalamus does not alter waking in the cat', *Brain Research* 539, pp. 287-303.

Desmedt, J.E. and Tomberg, C. (1989), 'Mapping early somatosensory evoked potentials in selective attention: critical evaluation of control conditions used for titrating by difference the cognitive P30, P40, P100 and N140', *Electroencephalography and Clinical Neurophysiology* 74, pp. 321-346.

Diehl, A.E. (1989), 'Human performance aspects of aircraft accidents', in Jensen, R.S. (ed.) *Aviation Psychology*, Gower Technical, Aldershot, pp. 378-403.

Diehl, A. (1991), 'The effectiveness of training programs for preventing aircrew error', in Jensen, R.S. (ed.) *Proceedings of the Sixth International Symposium on Aviation Psychology*, pp. 640-655.

Dinges, D.F. and Graeber, R.C. (1989), 'Crew fatigue monitoring', *Flight Safety Digest* 8(10), pp. 65-75.

Dolgin, D.L. and Gibb, G.D. (1989), 'Personality assessment in aviator selection', in Jensen, R.S. (ed.) *Aviation Psychology*, Gower Technical, Aldershot, pp. 288-320.

Duffy, E. (1972), 'Activation', in Greenfield, N.S. and Sternbach, R.A. (eds.) *Handbook of Psychophysiology*, Holt, New York, pp. 577-622.

Easterbrook, J.A. (1959), 'The effect of emotion on cue utilization and the organization of behavior', *Psychological Review* 66, pp. 183-201.

Elliott, R. (1964), 'Physiological activity and performance: a comparison of kindergarten children with young adults', *Psychological Monographs* 78(10, whole No. 587), pp. 1-33.

Emerson, T. and Reising, J. (1991), 'The effect of adaptive function allocation on the cockpit design paradigm', in Jensen, R.S. (ed.) *Proceedings of the Sixth International Symposium on Aviation Psychology,* pp. 114-121.

Ewing, D.J. (1988), 'Recent advances in the non-invasive investigation of diabetic autonomic neuropathy', in Bannister, R. (ed.) *Autonomic Failure*, 2nd edition, Oxford University Press, Oxford, pp. 667-689.

Eysenck, M.W. (1982), *Attention and Arousal*, Springer-Verlag, Berlin.

Fakhar, S., Vallet, M., Olivier, D. and Baez, D. (1991), 'La posture corporelle comme indicateur de fatigue du conducteur', in Vallet, M. (ed.) *Le maintien de la Vigilance dans les Transports*, L'Institut National de Recherche sur les Transports et leur Sécurité, Caen, pp. 151-157.

Fakhar, S., Vallet, M., Olivier, D. and Baez, D. (1992), 'Effet du bruit des vibrations sur la vigilance des conducteurs de vehicules legers en situation de conduite monotone', *Fiche Resultat Rapport Inrets* No.153, Laboratoire Energie Nuisances, Bron.

Feggeter, A.J.W. (1982), 'A method for investigating human factor aspects of aircraft accidents and incidents', *Ergonomics* 25, pp. 1065-1075.

Feggeter, A.J.W. (1991), 'The development of an intelligent human factors data base as an aid for the investigation of aircraft accidents', in Jensen, R.S. (ed.) *Proceedings of the Sixth International Symposium on Aviation Psychology,* pp. 624-629.

Fisk, A.D. and Schneider, W. (1981), 'Control and automatic processing during tasks requiring sustained attention: a new approach to vigilance', *Human Factors* 23, pp. 737-750.

Fouillot, J.-P., Coblentz, A., Cabon, P., Mollard, R. and Speyer, J.-J. (1991), 'Étude du niveau d'éveil des équipages au cours de vols long courriers', in Vallet, M. (ed.) *Le maintien de la Vigilance dans les Transports*, L'Institut National de Recherche sur les Transports et leur Sécurité, Caen, pp. 253-280.

Foushee, H.C. and Helmreich, R.L. (1988), 'Group interaction and flight crew performance', in Wiener, E.L. and Nagel, D.C. (eds.) *Human Factors in Aviation*, Academic Press, London, pp. 189-227.

Frankenhaeuser, M., Nordheden, B., Myrsten, A.-L. and Post, B. (1971), 'Psychophysiological reactions to understimulation and overstimulation', *Acta Psychologica* 35, pp. 298-308.

Frankenhaeuser, M., Lundberg, U., Fredrikson, M., Melin, B., Tuomisto, M. and Myrsten, A.-L. (1989), 'Stress on and off the job as related to sex and occupational status in white-collar workers', *Journal of Organizational Behaviour* 10, pp. 321-346.

Freeman, C. and Simmon, D.A. (1991), 'Taxonomy of crew resource management: information processing domain', in Jensen, R.S. (ed.) *Proceedings of the Sixth International Symposium on Aviation Psychology*, pp. 391-397.

Frone, M.R. and McFarlin, D.B. (1989), 'Chronic occupational stressors, self-focussed attention and well-being: testing a cybernetic model of stress', *Journal of Applied Psychology* 74, pp. 876-883.

Fusilier, M.R., Ganster, D.C. and Mayes, B.T. (1987), 'Effects of social support, role stress, and locus of control on health', *Journal of Management* 13, pp. 517-528.

Galinsky, T.L., Warm, J.S., Dember, W.N. and Weiler, E.M. (1990), 'Sensory alternation and vigilance performance: the role of pathway inhibition', *Human Factors* 32, pp. 717-728.

Gander, P.H. and Graeber, R.C. (1987), 'Sleep in pilots flying short-haul commercial schedules', *Ergonomics* 30, pp. 1365-1377.

Ganster, D.C. (1986), 'Type A Behavior and occupational stress', *Journal of Organizational Behavior Management* 8, pp. 61-84.

Geare, A.J. (1989), 'Job stress: boon as well as a bane', *Employee Relations* 11, pp. 1-26.

Geddes, N.D. (1991), 'Automatic display management using dynamic plans and events', in Jensen, R.S. (ed.) *Proceedings of the Sixth International Symposium on Aviation Psychology*, pp. 90-95.

Gibson, R.H. and Wilhelm, J. (1989), 'Managerial leadership assessment: personality correlates of and sex differences in ratings by leaders, peers, and followers', in Jensen, R.S. (ed.) *Proceedings of the Fifth International Symposium on Aviation Psychology*, pp. 682-685.

Glass, D.C. and Singer, J.E. (1972), *Urban stress. Experiments on noise and social stressors*, Academic Press, New York.

Glotzbach, S.F. and Heller, H.C. (1989), 'Thermoregulation', in Kryger M.H., Roth, T. and Dement, W.C. (eds.) *Principles and practice of sleep medicine*, WB Saunders, Philadelphia, pp. 300-310.

Graeber, R.C. (1986), (ed.) 'Sleep and wakefulness in international aircrews', *Aviation, Space and Environmental Medicine*, 57 (12, supplement), pp. B1-B64.

Graeber, R.C. (1988), 'Aircrew fatigue and circadian rhythmicity', in Wiener, E.L. and Nagel, D.C. (eds.) *Human Factors in Aviation*, Academic Press, London, pp. 305-344.

Green, R.G. (1985), 'Stress and accidents', *Aviation, Space and Environmental Medicine* 156, pp. 638-641.

Green, R.G., Muir, H., James, M., Gradwell, D. and Green, R.L. (1991), *Human factors for pilots*, Aldershot, Avebury Technical.

Greene, R.L., Kille, S.E. and Hogan, F.A. (1979), 'Electrodermal measures of attention and effort to stimulus onset and offset for intramodal and intermodal tasks', *Perceptual and Motor Skills* 48, pp. 411-418.

Gregorich, S.E., Helmreich, R.L. and Wilhelm, J.A. (1990), 'The structure of cockpit management attitudes', *Journal of Applied Psychology* 75, pp. 682-690.

Grossman, P., Stemmler, G. and Meinhardt, E. (1990), 'Paced sinus arrhythmia as an index of cardiac parasympathetic tone during varying behavioral tasks', *Psychophysiology* 27, pp. 404-416.

Gunn, W.H. (1991), 'Airline deregulation: impact on human factors', in Jensen, R.S. (ed.) *Proceedings of the Sixth International Symposium on Aviation Psychology,* pp. 662-667.

Habsheim (1990), 'Crew blamed for Habsheim air show A320 crash'. *Flight International* No.4197, 137, pp. 5.

Haider, M., Spong, P. and Lindsley, D.B. (1964), 'Attention, vigilance and cortical evoked-potentials in humans', *Science* 145, 180-182.

Halligan, P.W. and Marshall, J.C. (1991), 'Spatial compression in visual neglect: a case study', *Cortex* 27, pp. 623-629.

Hancock, P.A. (1991), 'On operator strategic behaviour', in Jensen, R.S. (ed.) *Proceedings of the Sixth International Symposium on Aviation Psychology,* pp. 999-1007.

Hancock, P.A. and Warm, J.S. (1989), 'A dynamic model of stress and sustained attention', *Human Factors 31*, pp. 519-537.

Hansen, K.A. and Duck, S.C. (1983), 'Teledyne sleep sentry: evaluation in pediatric patients for detection of nocturnal hypoglycemia'. *Diabetes Care* 6, pp. 579-600.

Hansman, R.J., Hahn, E. and Midkiff, A. (1991), 'Impact of data link on flight crew situational awareness', in Jensen, R.S. (ed.) *Proceedings of the Sixth International Symposium on Aviation Psychology,* p. 551.

Hawkes, G.R., Meighan, T.W. and Alluisi, E.A. (1964), 'Vigilance in complex task situations', *Journal of Psychology* 58, pp. 223-236.

Hawkins, F.H. (1987), *Human factors in flight*, Gower, Vermont.

Haworth, N.L. and Vulcan, P. (1991), *Testing of commercially available fatigue monitors*, Report No.15, Monash University Accident Research

Centre/ Australian Road Research Board.

Hayward, B. and Alston, N. (1991), 'Team building following a pilot labour dispute: extending the CRM envelope', in Jensen, R.S. (ed.) *Proceedings of the Sixth International Symposium on Aviation Psychology*, pp. 668-673.

Helmreich, R.L. (1991), 'Strategies for the study of flightcrew behaviour', in Jensen, R.S. (ed.) *Proceedings of the Sixth International Symposium on Aviation Psychology*, pp. 338-343.

Helmreich, R.L., Foushee, H.C., Benson, R. and Russini, W. (1986), 'Cockpit resource management: exploring the attitude-performance linkage', *Aviation, Space and Environmental Medicine* 57, pp. 1198-1200.

Helmreich, R.L. and Wilhelm, J.A. (1989), 'When training boomerangs: negative outcomes associated with cockpit resource management programs', in Jensen, R.S. (ed.) *Proceedings of the Fifth International Symposium on Aviation Psychology*, pp. 692-697.

Helmreich, R.L., Chidester, T.R., Foushee, H.C., Gregorich, S. and Wilhelm, J.A. (1990), 'How effective is cockpit resource management training?', *Flight Safety Digest* 9(5), pp. 1-17.

Hill, B.L. (1990), 'A320 operations under the microscope', *Aircraft & Aerospace* 70(8), pp. 14-15.

Hillyard, S.A. and Hansen, J.C. (1986), 'Attention: electrophysiological approaches', in Coles, M.G.H., Donchin, E. and Porges, S.W. (eds.) *Psychophysiology; Systems, Processes, and Applications*, Amsterdam, Elsevier, pp. 227-243.

Hjemdahl, P., Freyschuss, U., Juhlin-Dannfelt, A. and Linde, B. (1984) 'Differentiated sympathetic activation during mental stress evoked by the Stroop test', *Acta Physiologica Scandinavica* Suppl. 527, pp. 25-29.

Holt, G.W., Taylor, W.F. and Carter, E.T. (1985), 'Airline pilot disability: the continued experience of a major US airline', *Aviation, Space and Environmental Medicine* 56, pp. 939-944.

Hopkins, H. (1987), 'The state of the art. Flight-test Airbus A320', *Flight International* No.4092, 132, pp. 23-27.

Houston, R. (1990), 'Resource management in the cockpit', *Flight Safety Digest* 9(7), pp. 70-74.

Hovanitz, C.A., Chin, K. and Warm, J.S. (1989), 'Complexities in life stress-dysfunction relationships: a case in point-tension headache', *Journal of Behavioural Medicine* 12, pp. 55-75.

Howard, J.H., Cunningham, D.A. and Rechnitzer, P.A. (1986), 'Role ambiguity, type A behaviour and job satisfaction: moderating effects on cardiovascular and biochemical responses associated with coronary risk', *Journal of Applied Psychology* 71, pp. 95-101.

Hughes, D. (1989), 'Glass cockpit study reveals human factors problems', *Aviation Week and Space Technology* 6, pp. 32-34.

Hull, C.L. (1943), *Principles of behaviour*, Appleton-Century-Crofts, New York.

Inhoff, A.W., Rafal, R.D. and Posner, M.J. (1992), 'Bimodal extinction without cross-modal extinction', *Journal of Neurology, Neurosurgery and Psychiatry* 55, pp. 36-39.

Irwin, C. (1991), 'The impact of initial and recurrent cockpit resource management training on attitudes', in Jensen, R.S. (ed.) *Proceedings of the Sixth International Symposium on Aviation Psychology*, pp. 344-349.

James, M., McClumpha, A., Green, R., Wilson, P. and Belyavin, A. (1991), 'Pilot attitudes to cockpit automation', in Jensen, R.S. (ed.) *Proceedings of the Sixth International Symposium on Aviation Psychology*, pp. 192-197.

Jennings, J.R. (1986), 'Bodily changes during attending', in Coles, M.G.H., Donchin, E. and Porges, S.W. (eds.) *Psychophysiology; Systems, Processes, and Applications*, Elsevier, Amsterdam, pp.268-289.

Jensen, R.S. and Biegelski, C.S. (1989), 'Cockpit resource management', in Jensen, R.S. (ed.) *Aviation Psychology*, Gower Technical, Aldershot, pp. 176-210.

Jewett, D.L. and Williston, J.S. (1971), 'Auditory-evoked far fields averaged from the scalp of humans', *Brain* 94, pp. 681-696.

Johnston, N. (1991), 'Organizational factors in human factors accident investigations', in Jensen, R.S. (ed.) *Proceedings of the Sixth International Symposium on Aviation Psychology*, pp. 668-673.

Jorna, P.G.M. (1991), 'Heart rate variability as an index for pilot workload', in Jensen, R.S. (ed.) *Proceedings of the Sixth International Symposium on Aviation Psychology*, pp. 746-751.

Judge, C.L. (1991), 'Lessons learned about information management within the pilot's associate program', in Jensen, R.S. (ed.) *Proceedings of the Sixth International Symposium on Aviation Psychology*, pp. 85-89.

Kahn, A. (1991), 'Behavioural analysis of management actions in aircraft accidents', in Jensen, R.S. (ed.) *Proceedings of the Sixth International Symposium on Aviation Psychology*, pp. 674-678.

Kahn, R.L., Wolfe, D.M., Quinn, R.P., Snoek, J.D. and Rosenthal, R.A. (1964), *Organizational stress: studies in role conflict and ambiguity*, Wiley, New York.

Kahneman, D., Tursky, B., Shapiro, D. and Crider, A. (1969), 'Pupillary, heart rate, and skin resistance changes during a mental task', *Journal of Experimental Psychology* 79, pp. 164-167.

Kanki, B.G. (1991), 'Session overview: information transfer and crew performance', in Jensen, R.S. (ed.) *Proceedings of the Sixth International Symposium on Aviation Psychology*, pp. 247-252.

Kanki, B.G., Greaud, V.A. and Irwin, C.M. (1989), 'Communication variations and aircrew performance', in Jensen, R.S. (ed.) *Proceedings of the Fifth International Symposium on Aviation Psychology*, pp. 419-424.

Kanki, B.G., Lozito, S. and Foushee, H.C. (1989), 'Communication indices of crew coordination', *Aviation Space and Environmental Medicine* 60, pp. 56-60.

Kanki, B.G., Palmer, M. and Veinott, E. (1991), 'Communication variations related to leader personality', in Jensen, R.S. (ed.) *Proceedings of the Sixth International Symposium on Aviation Psychology*, pp. 253-259.

Kantowitz, B.H. and Casper, P.A. (1988), 'Human workload in aviation'. in Wiener, E.L. and Nagel, D.C. (eds.) *Human Factors in Aviation*, Academic Press, London, pp. 157-187.

Karlins, M., Koh, F. and McCully, L. (1989), 'The spousal factor in pilot stress', *Aviation Space and Environmental Medicine* 60, pp. 1112-1115.

Kaufmann, G.M. and Beehr, T.A. (1986), 'Interactions between job stressors and social support: some counterintuitive results', *Journal of Applied Psychology* 71, pp. 522-526.

Kemery, E.R., Bedeian, A.G., Mossholder, K.W. and Touliatos, J. (1985), 'Outcome of role stress: a multisample constructive replication', *Academy of Management Journal* 28, pp. 363-375.

Kennedy, J.L. and Travis, R.C. (1947), 'Prediction of speed of performance by muscle action potentials', *Science* 105, pp. 410-411.

Koelega, H.S., Brinkman, J.A., Hendriks, L. and Verbaten, M.N. (1989), 'Processing demands, effort, and individual differences in four different vigilance tasks', *Human Factors* 31, pp. 45-62.

References

Komich, J.N. (1991), 'CRM scenario development: the next generation', in Jensen, R.S. (ed.) *Proceedings of the Sixth International Symposium on Aviation Psychology*, pp. 53-59.

Koonce, J.M. (1989), 'Another look at aircraft accident statistics', in Jensen, R.S. (ed.) *Proceedings of the Fifth International Symposium on Aviation Psychology*, pp. 866-871.

Lautman, L.G. and Gallimore, P.L. (1989), 'Control of crew-caused accidents', *Flight Safety Digest* 8(10), pp. 76-88.

Learmount, D. (1992), 'Human factors', *Flight International* No.4238, 139, pp. 30-33.

Leedom, D.K. (1991), 'Aircrew coordination for army helicopters: research overview', in Jensen, R.S. (ed.) *Proceedings of the Sixth International Symposium on Aviation Psychology*, pp. 284-289.

Levine, P. (1986), 'Stress', in Coles, M.G.H., Donchin, E. and Porges, S.W. (eds.) *Psychophysiology; Systems, Processes, and Applications*, Elsevier, Amsterdam, pp. 331-353.

Lindberg, C.A. (1953), *The spirit of St. Louis*, Charles Scribner's Sons, New York.

Little, L.F., Gaffney, I.C., Rosen, K.H. and Bender, M.M. (1990), 'Corporate instability is related to airline pilot's stress symptoms', *Aviation Space and Environmental Medicine* 61, pp. 977-982.

Loeb, M. and Alluisi, E.A. (1984), 'Theories of vigilance', in Warm, J.S. (ed.) *Sustained Attention in Human Performance*, Wiley, Chichester, pp. 179-205.

Logan, A.L. and Braune, R.J. (1991), 'The utilization of the aviation safety reporting system: a case study in pilot fatigue', in Jensen, R.S. (ed.) *Proceedings of the Sixth International Symposium on Aviation Psychology*, pp. 793-798.

Mackie, R.R. and Wylie, C.D. (1991), 'Countermeasures to loss of alertness in truck drivers. Theoretical and practical considerations', in Vallet, M. (ed.) *Le maintien de la Vigilance dans les Transports*, L'Institut National de Recherche sur les Transports et leur Sécurité: Caen, pp. 113-141.

Mackworth, J.F. (1965), 'The effect of amphetamine on the detectability of signals in a vigilance task', *Canadian Journal of Psychology* 19, pp. 104-110.

Mackworth, J.F. (1970), *Vigilance and attention*, Penguin Books, Baltimore.

Maher, J.W. (1989), 'Beyond CRM to decisional heuristics: an airline generated model to examine accidents and incidents caused by crew errors in deciding', in Jensen, R.S. (ed.) *Proceedings of the Fifth International Symposium on Aviation Psychology*, pp. 439-444.

Malmo, R.B. (1959), 'Activation: a neuropsychological dimension', *Psychological Review* 66, pp. 367-386.

Margerison, C.J., Davies, R.V. and McCann, D.J. (1986), 'Team management of the flight deck', *Leadership and Organization Development Journal* 7, pp. 3-26.

Marino, K.E. and White, S.E. (1985), 'Departmental structure, locus of control, and job stress: the effect of a moderator', *Journal of Applied Psychology* 4, pp. 782-784.

Mason, J.W. (1971), 'A re-evaluation of the concept of 'non-specificity' in stress theory', *Journal of Psychiatric Research* 8, pp. 323-333.

Masson, P. (1991), 'Le maintien de l'éveil des conducteurs de trains', in Vallet, M. (ed.) *Le maintien de la Vigilance dans les Transports*, L'Institut National de Recherche sur les Transports et leur Sécurité: Caen, pp. 59-64.

Matteson, M.T. and Ivancevich, J.M. (1987), *Controlling work stress. Effective human resource and management strategies*, Jossey-Bass, San Francisco.

Mayes, B.T. and Ganster, D.C. (1988), 'Exit and voice: a test of hypotheses based on fight/flight responses to job stress', *Journal of Organizational Behaviour* 9, pp. 199-216.

McLeod, J.G. and Tuck, R.R. (1987), 'Disorders of the autonomic nervous system: Part 2 Investigation and treatment', *Annals of Neurology* 21, pp. 519-529.

Mennemeier, M., Wertman, E. and Heilman, K.M. (1992), 'Neglect of near peripersonal space', *Brain* 115, pp. 37-50.

Morris, M., Lack, L. and Dawson, D. (1989), 'Sleep-onset insomniacs have delayed temperature rhythms', *Sleep* 13, pp. 1-14.

Morrison, J.G., Gluckman, J.P. and Deaton, J.E. (1991), 'Human performance in complex task environment: a basis for the application of adaptive automation', in Jensen, R.S. (ed.) *Proceedings of the Sixth International Symposium on Aviation Psychology*, pp. 96-101.

Mortimer, R.G. (1991), 'Some factors associated with pilot age in general aviation crashes', in Jensen, R.S. (ed.) *Proceedings of the Sixth International Symposium on Aviation Psychology*, pp. 770-775.

Mosier, K. (1991), 'Expert decision making strategies', in Jensen, R.S. (ed.) *Proceedings of the Sixth International Symposium on Aviation*

Psychology, pp. 266-271.
Mosier-O'Neill, K.L. (1989), 'A contextual analysis of pilot decision making', in Jensen, R.S. (ed.) *Proceedings of the Fifth International Symposium on Aviation Psychology*, pp. 371-376.
Moss Kanter, R. (1989), *When Giants Learn to Dance*, Unwin, London.
Moxon, J. (1991), 'Airbus offers autothrottle option', *Flight International* No.4265, 139, pp. 20.
Murphy, L.R. (1986), 'A review of organizational stress management research: methodological considerations', *Journal of Organizational Behavior Management* 8, pp. 215-227.
Nagel, D.C. (1988) 'Human error in aviation operations', in Wiener, E.L. and Nagel, D.C. (eds.), *Human Factors in Aviation*, Academic Press, London, pp. 263-303.
Neiss, R. (1988), 'Reconceptualising arousal: psychological states in motor performance', *Psychological Bulletin* 103, pp. 345-366.
Neiss, R. (1990), 'Ending arousal's reign of terror: a reply to Anderson', *Psychological Bulletin* 107, pp. 101-105.
Norman, S., Billings, C.E., Nagel, D., Palmer, E., Wiener, E.L. and Woods, D.D. (1988), 'Aircraft automation philosophy: a source document', for NASA/Industry/FAA workshop, *Flight Deck Automation: Promises and Realities*, National Aeronautics and Space Administration, Ames Research Center, California.
Norris, G. (1991), 'Lauda thrust-reverser under suspicion', *Flight International* No.4271, 139, pp. 16-17.
O'Hare, D. and Roscoe, S. (1990), *Flightdeck Performance. The Human Factor*, Iowa State University Press, Ames.
Orlady, H.W. (1989), 'Training for advanced cockpit technology aircraft', *Flight Safety Digest* 8(10), pp. 50-54.
Orlady, H.W. (1990), 'Todays professional airline pilot: all the old skills and more', *Flight Safety Digest* 9(6), pp. 1-6.
Palmer, E. and Degani, A. (1991), 'Electronic checklists: evaluation of two levels of automation', in Jensen, R.S. (ed.) *Proceedings of the Sixth International Symposium on Aviation Psychology*, pp. 178-183.
Parasuraman, R. (1979), 'Memory load and event rate control sensitivity decrements in sustained attention', *Science* 205, pp. 924-927.
Parasuraman, R. (1984), 'The psychobiology of sustained attention', in Warm, J.S. (ed.) *Sustained attention in human performance*, Wiley, Chichester, pp. 61-101.
Parasuraman, R., Bahri, T., Molloy, R. and Singh, I. (1991), 'Effects of

shifts in the level of automation on operator performance', in Jensen, R.S. (ed.) *Proceedings of the Sixth International Symposium on Aviation Psychology*, pp. 102-107.

Parasuraman, R. and Davies, D.R. (1976), 'Decision theory analysis of response latencies in vigilance', *Journal of Experimental Psychology: Human Perception and Performance* 2, pp. 578-590.

Pardo, J.V., Fox, P.T. and Raichle, M.E. (1991), 'Localization of a human system for sustained attention by positron emission tomography', *Nature* 349, pp. 61-64.

Pardo, J.V., Pardo, P.J., Janer, K.W. and Raichle, M.E. (1990), 'The anterior cingulate cortex mediates processing selection in the Stroop attentional conflict paradigm', *Proceedings National Academy Science* 87, pp. 256-259.

Pawlik, E.A., Simon, R. and Dunn, D. (1991), 'Aircrew coordination for army helicopters: improved procedures for accident investigation', in Jensen, R.S. (ed.) *Proceedings of the Sixth International Symposium on Aviation Psychology*, pp. 320-325.

Persson, B., Andersson, O.K., Hjemdahl, P., Wysocki, M., Agerwall, S. and Wallin, G. (1989), 'Adrenaline infusion in man increases muscle sympathetic nerve activity and noradrenaline overflow in plasma', *Journal of Hypertension* 7, pp. 747-756.

Peter, J.H., Cassel, W., Ehrig, B., Faust, M., Fuchs, E., Langanke, P., Meinzer, K. and Pfaff, U. (1990a), 'Occupational performance of a paced secondary task under conditions of sensory deprivation. I. Heart rate changes in train drivers as a result of monotony', *European Journal of Applied Physiology* 60, pp. 309-314.

Peter, J.H., Cassel, W., Ehrig, B., Faust, M., Fuchs, E., Langanke, P., Meinzer, K. and Pfaff, U. (1990b), 'Occupational performance of a paced secondary task under conditions of sensory deprivation. II. The influence of professional training', *European Journal of Applied Physiology* 60, pp. 315-320.

Petit, C. and Tarrière, C. (1991), 'Effets synergiques du bruit, des vibrations et de la chaleur sur la vigilance du conducteur', in Vallet, M. (ed.) *Le maintien de la Vigilance dans les Transports*, L'Institut National de Recherche sur les Transports et leur Sécurité, Caen, pp. 171-183.

Pew, R.W. (1986), 'Human performance issues in the design of future air force systems', *Aviation Space and Environmental Medicine* 57(10 Suppl), pp. A78-A82.

Poe, G.R., Suyenobu, B.Y., Bolstad, C.A., Endsley, M.R. and Sterman, M.B. (1991), 'EEG correlates of critical decision making in computer simulated combat', in Jensen, R.S. (ed.) *Proceedings of the Sixth International Symposium on Aviation Psychology*, pp. 758-763.

Pope, J.A. (1991), 'When a rejected takeoff goes bad', *Flight Safety Digest* 10(2), pp. 1-16.

Posner, M.I., Petersen, S.E., Fox, P.T. and Raichle, M.E. (1988), 'Localization of cognitive operations in the human brain', *Science* 249, pp. 1627-1631.

Poulton, E.C. (1979), 'Composite model for human performance in continuous noise', *Psychological Review* 86, pp. 361-375.

Povenmire, H.K., Rockway, M.R., Bunecke, J.L. and Patton, M.W. (1989), 'Cockpit resource management skills enhance combat mission performance in a B-52 simulator', in Jensen, R.S. (ed.) *Proceedings of the Fifth International Symposium on Aviation Psychology*, pp. 489-494.

Predmore, S. (1991), 'Microcoding of communications in accident investigation: crew coordination in United Flight 811 and United 232', in Jensen, R.S. (ed.) *Proceedings of the Sixth International Symposium on Aviation Psychology*, pp. 350-355.

Pribram, K.H. and McGuinness, D. (1975), 'Arousal, activation, and effort in the control of attention', *Psychological Review* 82, pp 116-149.

Quick, J.D., Horn, R.S. and Quick, J.C. (1986), 'Health consequences of stress', *Journal of Organizational Behavior Management* 8, pp. 19-36.

Reason, J. (1991), 'Identifying the latent causes of aircraft accidents before and after the event', *International Society of Air Safety Investigators*, 22nd Annual Seminar, Canberra.

Rogers, W. (1991), 'Information management:assessing the demand for information', in Jensen, R.S. (ed.) *Proceedings of the Sixth International Symposium on Aviation Psychology*, pp. 66-71.

Roscoe, S.N. (1991), 'Simulator qualification: just as phony as it can be', in Jensen, R.S. (ed.) *Proceedings of the Sixth International Symposium on Aviation Psychology*, pp. 868-872.

Roscoe, A.H. (1992), 'Workload in the glass cockpit', *Flight Safety Digest* April, pp. 1-9.

Sanders, A.F. (1983), 'Towards a model of stress and human performance', *Acta Psychologica* 53, pp. 61-97.

Schaubroeck, J., Cotton, J.L. and Jennings, K.R. (1989), 'Antecedents and consequences of role stress: a covariance structure analysis', *Journal of Organizational Behaviour* 10, pp. 35-85.

Schnore, M.M. (1959), 'Individual patterns of physiological activity as a function of task differences and degree of arousal', *Journal of Experimental Psychology* 58, pp. 117-128.

Schwartz, D. (1989), 'Training for situational awareness', in Jensen, R.S. (ed.) *Proceedings of the Fifth International Symposium on Aviation Psychology*, pp. 44-54.

Schwartz, G.E., Weinberger, D.A. and Singer, J.A. (1981), 'Cardiovascular differentiation of happiness, sadness, anger, and fear following imagery and exercise', *Psychosomatic Medicine* 43, pp. 343-364.

Sellards, R. (1989), 'Testing for potential problem pilots and human error in the cockpit', in Jensen, R.S. (ed.) *Proceedings of the Fifth International Symposium on Aviation Psychology*, pp. 582-587.

Selye, H. (1956), *The Stress of Life*, McGraw-Hill, New York.

Sharpless, S. and Jasper, H. (1956), 'Habituation of the arousal reaction', *Brain* 79, pp. 655-680.

Simmel, E.C., Cerkovnik, M. and McCarthy, J.E. (1989), 'Sources of stress affecting pilot judgment', *Aviation Space and Environmental Medicine* 60, pp. 53-55.

Sloan, S.J. and Cooper, C.L. (1985), 'The impact of life events on pilots: an extension of Alkov's approach', *Aviation Space and Environmental Medicine* 56, pp. 845-859.

Speyer, J.J., Monteil, C., Blomberg, R.D. and Fouillot, J.P. (1991), 'Impact of new technology on operational interface: from design aims to flight evaluation and measurement', in Vallet, M. (ed.) *Le maintien de la Vigilance dans les Transports*, L'Institut National de Recherche sur les Transports et leur Sécurité, Caen, pp. 191-252.

Spyer, K.M. (1988), 'Central nervous system control of the cardiovascular system', in Bannister, R. (ed.) *Autonomic Failure*, Oxford University Press, Oxford, pp. 56-79.

Sterman, M.B. and Olff, M. (1991), 'Topographic EEG correlates of perceptual defensiveness', in Jensen, R.S. (ed.) *Proceedings of the Sixth International Symposium on Aviation Psychology*, pp. 764-769.

Stragisher, G.W. (1991), 'Teaching an old dog new tricks: concepts, schemata and metacognition in pilot training and education training', in Jensen. R.S. (ed.) *Proceedings of the Sixth International Symposium on Aviation Psychology*, pp. 958-963.

Swezey, R., Llaneras, R., Prince, C. and Salas, E. (1991), 'Instructional strategy for aircrew coordination training', in Jensen, R.S. (ed) *Proceedings of the Sixth International Symposium on Aviation Psychology*, pp. 302-307.

Taggart, W.R. and Butler, R.E. (1989), 'CRM validation program', in Jensen, R.S. (ed.) *Proceedings of the Fifth International Symposium on Aviation Psychology*, pp. 468-481.

Tarrière, C., Hartemann, E., Sfez, E., Chaput, D. and Petit-Poilvert, C. (1988), 'Some ergonomic features of the driver-vehicle-environment interface', *The Engineering Society for Advanced Mobility Land Sea Air and Space*, No 885051, Washington.

Taylor, R.M. (1989), 'Crew systems design: some defence, psychological futures', in Jensen, R.S. (ed.) *Aviation Psychology*, Gower Technical, Aldershot, pp. 38-65.

Tenney, Y.J., Adams, M.J., Pew, R.W., Huggins, A.W.F. and Rogers W.H. (1992), 'A principled approach to the measurement of situation awareness in commercial aviation', *NASA Contractor Report 4451* Langley Research Center, National Aeronautics and Space Administration.

Tetrick, L.E. and LaRocco, J.M. (1987), 'Understanding, prediction, and control as moderators of the relationship between perceived stress, satisfaction, and psychological well-being', *Journal of Applied Psychology* 72, pp. 538-543.

Thackray, R.I. (1981), 'The stress of boredom and monotony: a consideration of the evidence', *Psychosomatic Medicine* 43, pp. 165-176.

Thomas, M. (1989), *Managing Pilot Stress*, Macmillan, New York.

Thomas, M. (1991), 'Stress management for the third revolution aviator', in Jensen, R.S. (ed.) *Proceedings of the Sixth International Symposium on Aviation Psychology*, pp. 38-43.

Thorsden, M.L. and Klein, G.A. (1991), 'Training implications of a team decision model', in Jensen, R.S. (ed.) *Proceedings of the Sixth International Symposium on Aviation Psychology*, pp. 296-301.

Tsang, P.S. and Vidulich, M.A. (1989), 'Cognitive demands of automation in aviation', in Jensen, R.S. (ed.) *Aviation Psychology* Gower Technical, Aldershot, pp. 66-96

Tucker, D.M. and Williamson, P.A. (1984), 'Asymmetric neural control systems in human self-regulation', *Psychological Review* 91, pp. 185-215.

Tulen, J.H., Moleman, P., Van Steenis, H.G. and Boomsma, F. (1989), 'Characterization of stress reactions to the Stroop Color Word Test', *Pharmacology, Biochemistry and Behavior* 32, pp. 9-15.

Vallet, M. (1991), 'Les dispositifs de maintien de la vigilance des conducteurs de voitures', in Vallet, M. (ed.) *Le maintien de la Vigilance dans les Transports*, L'Institut National de Recherche sur les Transports et leur Sécurité, Caen, pp. 13-21.

Vanderwolf, C.H. (1992), 'The electrocorticogram in relation to physiology and behavior: a new analysis', *Electroencephalography and Clinical Neurophysiology* 82, pp. 165-175.

Vanderwolf, C.H. and Robinson, T.E. (1981) 'Reticulo-cortical activity and behavior: a critique of the arousal theory and a new synthesis', *The Behavioral and Brain Sciences* 4, pp. 459-514.

Walter, W.G., Cooper, R., Aldridge, V.J., McCallum, W.C. and Winter A.L. (1964), 'Contingent negative variation: an electrical sign of sensorimotor association and expectancy in the human brain', *Nature* 203, pp. 380-384.

Warm, J.S. (1984), 'An introduction to vigilance', in Warm, J.S. (ed.) *Sustained attention in human performance*, Wiley, Chichester, pp. 1-14.

Warren, R.A., Hudy, J.J. and Gratzinger, P. (1991), 'The myth of the adventuresome aviator', in Jensen, R.S. (ed.) *Proceedings of the Sixth International Symposium on Aviation Psychology*, pp. 700-705.

Warwick, G. (1989), 'Future trends and developments', in Middleton D.H. (ed.) *Avionic Systems*, Longman, Essex, pp. 248-257.

Wiener, E.L. (1988), 'Cockpit automation', in Wiener, E.L. and Nagel D.C. (eds.) *Human Factors in Aviation*, Academic Press, London, pp. 433-461.

Wiener, E.L. (1989), 'Human factors of advanced technology ("glass cockpit") transport aircraft', *NASA Contractor Report 177528*, Contract NCC2-377, National Aeronautics and Space Administration, Ames Research Center, California.

Wiener, E.L., Chidester, T.R., Kanki, B.G., Palmer, E.A., Curry, R.E. and Gregorich, S.E. (1991), 'The impact of cockpit automation on crew coordination and communication: 1. Overview, LOFT evaluations, error severity, and questionnaire data', *NASA Contractor Report* 177587, National Aeronautics and Space Administration, Ames Research Center, California.

Wiener, E.L. and Curry, R.E. (1980), 'Flight-deck automation: promises and problems', *Ergonomics* 23, pp. 995-1011.

Westerlund, E. (1991), 'The judgement styles model: a new tool for analysis and training', in Jensen, R.S. (ed.) *Proceedings of the Sixth International Symposium on Aviation Psychology,* pp. 1062-1067.

Wilson, A. (1989), 'Aircrew team management - the Australian experience', in Jensen, R.S. (ed.) *Proceedings of the Fifth International Symposium on Aviation Psychology,* pp. 462-467.

Yanowitch, R.E. (1977), 'Crew behaviour in accident causation'. *Aviation Space and Environmental Medicine* 48, pp. 918-921.

Yerkes, R.M. and Dodson, J.D. (1908), 'The relation of strength of stimulus to rapidity of habit-formation', *Journal of Comparative Neurology and Psychology* 18, pp. 459-482.

Yoss, R.E., Moyer, N.J. and Hollenhorst, R.W. (1970), 'Pupil size and spontaneous pupillary waves associated with alertness, drowsiness and sleep', *Neurology* 201, pp. 545-554.

Index

accident 8, 13-18, 23-27, 31-32, 36-46, 81, 100, 131, 137 139
activity-based 32-35
acute reactive stress 21, 28, 35, 49, 54, 57
adaptive automation 8, 12, 131
adrenalin 49, 74, 76-77, 97-98
adrenalin, measurement 97-98
Airbus 93, 100, 106
Airbus research 93, 100, 106
A320 14, 26-27, 42-43, 81, 100
A330/340 93, 100
A340 crew rotation 93
aircraft attentional system 109-110
Air Florida 40-41
Air France (Habsheim) 42
Air Inter (Strasbourg) 14, 26, 43
air traffic control 9-10, 15, 27, 50, 115, 124-125, 151
air traffic controller disease patterns 50
alerting (system) 2, 22, 103-104, 106 109-113, 117, 119-121, 124-125, 130, 132-133, 139, 148-152
alerting system classification 110
alertness displays 115
algorithm 106, 126, 131, 149
amphetamine 78
anaesthetist 12
anterior cingulate gyrus 70, 76

arousal 2, 21, 47, 51-56, 70-71, 75-76, 89, 91-93, 95-99, 102, 105, 120-122, 126, 141, 151
arousal (definition) 47, 51, 52
attention 59, 61-63, 68-70, 72-73, 80, 83, 89
attention and arousal system 73, 80, 89
attention definition 59
attention measurement 65, 70, 90, 92, 97
attention measuring system 103-109, 111-116
attention measuring system effectiveness 111-116
attentional channels 62
attitudes 32, 38, 140, 142-143
attitudes to automation 140, 142-143
automatic behaviour 107, 112-113, 115, 120
automation 2-3, 7-15, 18-22, 26-32, 34, 38-46, 54-55, 59, 81-82, 84-85, 100, 131, 139-145, 147-150
automation and CRM 29, 31, 38, 43-46, 141, 145
automation and complacency 11, 20
automation and peripheralisation 2-3, 7, 10, 12, 14-15, 18-19, 27-30, 39-46, 81, 85, 139

177

automation and role change 10, 55, 59
automation and vigilance 81, 82, 84, 85, 139, 148, 149
automation and workload 21-22, 55, 100
automation deficit 12, 22
automation, definition 7
automation philosophy 8-10, 26-27, 30, 144-145, 147
autonomy 130-131, 143-144
autonomy and alerting system 119-120, 124-125, 133
autopilot 9
Avianca 41-42
aviation alerting system 103, 109, 113

big picture 14
Boeing 2, 12-13, 17, 21, 24, 26, 29, 36, 41, 54, 109-110, 112-113, 115, 117, 140, 143-144, 150
Boeing alerting system 2, 109-110, 112-113, 117, 150
Boeing cockpit design 26, 29
Boeing information display 13
Boeing 737 24, 36, 40
Boeing 747 2, 14, 24, 41, 144, 155
Boeing 757 12, 21, 26, 54
Boeing 757 workload 21, 26, 115, 140, 143
boredom 12, 22, 147

causal-correlational issue 65-67, 77
cerebral blood flow 70
checklists and automation 11-12

climb phase 141, 150
clinical examples 68-71, 93, 122
CRM 31-46, 56, 85, 140-141, 145-147
CRM and accidents 31, 32, 36, 39-40
CRM and automation 44-46
CRM and communication 32, 34, 37, 41, 45
CRM and decision making 32, 41
CRM and error 29, 31, 37, 39
CRM as tangential solution 43, 146-147
CRM effectiveness 35-41, 43, 45-46, 85
CRM exposed crews 36, 45
CRM, future prospects 43, 145-147
CRM goals 32, 43, 46
CRM negative response 38
CRM origins 31
cognate state 14-15, 22, 40-41, 130, 147, 150
cognate state optimisation 130, 147, 150
cognitive stressors 56-57, 83, 120
command responsibility 32, 37, 41-42
communication 10, 12-13, 17-20, 22-24, 26-28, 32, 34, 37-38, 41-43, 46, 81, 93, 138, 147
communication and CRM 32, 34, 38
communication and errors 34
communication and personality 23
communication as accident factor

13, 17, 41-43, 93, 147
communication, crew/team factors 19-20, 34
communication effectiveness 13, 17-18, 34
communication, effects of automation 13, 34, 93
communication, effects of peripheralisation 10, 13, 18, 23-24, 43, 45-46, 81, 138
communication training 37-38
complacency 10-13, 17-20, 22-23, 27-30, 41-46, 81, 100, 114
complacency, boredom-panic 12
complacency and primary/secondary task inversion 11-12
complacency and vigilance 30, 81
complacency as error modulator 20-21, 120
complacency, definition 11
complacency, peripheralisation 10-13, 18, 20-21, 28, 81, 100
consequences of alerting system 150
continuum (arousal) 52, 60, 67
control systems and automation 8-10
coordination, crew 11-13, 24, 32, 37-38, 41, 142, 144, 146
coronary artery disease 51
cortex (cortical) 60, 68-72, 76-79, 93-94, 107
cortical systems and attention 68-70
costs 7-8, 30, 36, 38, 85, 111, 113, 132-133, 141, 149, 151-152, 156
costs and alerting system 111, 113, 131-133, 149, 151-152, 156
costs and automation 30, 85
costs of monitoring failure 151
crew duties 9, 148-150
crew rotation 93, 96, 101, 140 141, 145
crew rotation and arousal 141
crew training and automation 140, 142-144, 150
criterion shift 61, 79
critique (in communication) 34
cruise phase accidents 137-138
cruise phase management 9, 132, 138-147
cruise phase role 9, 138-139,
cruise phase workload 9, 138
cue utilisation 21, 64-65
culture 25, 121
culture, organisational 19, 24-25, 125

datalink (ATC) 15, 27
defence reaction 74
descent phase 10, 26, 39, 42, 43, 96, 115, 133, 137-138, 140-141, 146, 150
design, aircraft 7, 125, 141, 150
design intent 26, 114, 140
design, job 3, 103
design of alerting devices 105, 119-120, 125, 134, 150-151
design of man-machine interface 26, 99-100, 134
dipole 70
drive theory 64

economic(s) 7-9, 27, 30, 144
economic factors and automation

9, 27, 30
effectiveness of alerting system 111-113
effort and arousal 99, 102, 151
EEG and arousal 91-93
EEG and attention 70, 90-95, 101, 104, 106, 115, 123
EEG measurement 91-94, 98
endocrine 52-53, 65, 74, 77
environmental (stressors) 49, 54, 57, 69
error 10, 12, 15-18, 20-31, 34, 36-37, 39, 43-46, 81-82, 85, 100, 106, 132, 138, 149, 155
error and peripheralisation 10, 18, 27-29, 43-46, 81-82, 85
error and vigilance 81-82, 85, 149
error, human 15-17
error modulators 18-26, 138
error taxonomy 36
ERP 69-70
exchange theory 24
expectancy theory 61-62

fatigue 22-23, 93
faultless aircraft 13, 18, 27, 30, 41, 43, 81, 100, 138
fight or flight 74
filter theory 61-62
FMS 115
flight path, consequence of change 10, 39
flight path, optimal 8
flight phase 137-139
flight phase, accident incidence 137-138
flight phase, pattern of sweat output 127-129

flight phase, vigilance levels 138
followership 34
future of alerting system 130, 139, 149-150

good stress 52
GPWS 26, 42, 125

habituation 60-61
health and alerting systems 1, 121, 132-133
heart rate 49, 72-74, 93-96, 99-101, 123, 128
heart rate and arousal 95, 99
heart rate and vigilance 72-73, 99, 123
heart rate and workload 99-101
heart rate, measurement in aircraft 93, 96, 99-101, 128
heart rate variability 72, 99-100, 123
history of alerting systems 103
human-centred 8, 10, 12, 15, 26, 82, 130-133
human error (see error and accident)
human factors 8, 13-17, 23, 31, 36-46, 81, 100, 137-138, 150
human factors accidents 8, 13-17, 36-37, 39-46, 138
human factors experts 100, 150
hypertension 51
hypothalamus 74-77, 96-97, 123

ideal alerting system 119-121, 139, 148
incidents (see accidents) 11, 13-15, 18, 20, 22-23, 26-27, 29-30, 35-37, 138

Index

incidents, sleep related 109, 146
Indian Airlines 42
information access 14, 24-26, 41, 140,
information acquisition 11, 13-15, 22, 26, 83, 147
information in attentional systems 70, 74, 101, 110, 122-123, 126, 130, 147
information management 143, 147, 152
information processing 8, 13-15, 21-22, 24, 116, 147
information theory (vigilance) 60-62, 79
information transfer in CRM 32
insomniac 75
interface (man-machine) 1, 10-13, 26, 43, 89, 100, 116, 131
interface, human-hardware 10
interface, human-software 10
interface, human-procedural 10
inverted U-hypothesis 64-66
investigation, accident 17
investigation, safety 36

Kegworth 27
knowledge-based 32, 36

landing phase 9, 39, 128, 137-138, 141, 150
lay view of arousal 51
leader personality 23-24
leadership 19, 23-24, 34
leadership as error modulator 19, 23-24
leadership in CRM 34
learning models 60
life stress 21, 49, 53, 55, 57, 147

Likert 37
Lindberg 2
loop, arousal, open/closed 51-52
magneto encephalogram 107
management 25, 151-152
management, cockpit 8-9, 18, 21, 25, 31, 37, 43, 110, 130, 139-144, 149-153
management, flight 9, 140
management, flight operations 110, 143, 149
measurement of arousal 121, 122
metacognition 20
monitoring 1-3, 9, 12, 21-22, 29-30, 42, 48, 50-51, 54-56, 67, 79, 81-85, 93, 96, 103, 109, 112-117, 120-121, 124, 130-132, 137-153
monitoring CRM processes 32, 35-37
monitoring consequences 12, 21-22, 55, 81, 138-139
monitoring, faulty 42, 103, 134, 141
monitoring, measurement of 109, 112-116, 143
monitoring management 8, 67, 79, 85, 117, 12, 130-132, 137-153
monitoring processes 29-30, 51, 54, 56, 79, 81, 83-85, 120, 139
monitoring reserve 131-132, 149
monitoring role 9, 29-30, 48, 50, 55-56, 79, 81-85, 103, 109, 112, 132, 138, 142-153
movement (body part) 93, 98, 101, 103, 141

NASA 11, 17, 31, 37, 109, 146

neglect 69
neurology (ical) 68, 93
neurophysiology (ical) 65, 68-71
new error forms 132
noise and attention 61, 65, 91-92, 147
noise as a stressor 48, 53, 83
noise in measurement 89-91, 94, 96-98, 106, 123, 126, 129-130
noise, measurement types 89-90
non-routine events (see acute reactive stress)

occupational health 49, 132
organisation 24-25, 48-49, 54, 130, 139, 142, 150, 152
organisational culture 24-25, 33, 125
orienting response 63, 97
overarousal 65

parietal 69-70, 80
performance enhancement (arousal) 21, 53, 54, 120
peripheralisation 1, 3, 10, 12-15, 18-30, 38-46, 56, 81, 83-85, 100, 112, 114, 133, 139, 142-143, 149
peripheralisation and CRM 38, 41, 43-46
peripheralisation and monitoring 29-30, 139
peripheralisation, caused by alerting systems 112, 114
peripheralisation, definition 10
peripheralisation effects 10-15, 100, 139
peripheralisation error modulators 18-26, 142

peripheralisation reversal 132, 139, 142-143, 149
peripheralisation schema 27-28, 44-46, 82, 84
peripheralisation stressors 56, 83, 139
personality 19, 23-24, 32, 78, 121, 146
PET and arousal 70
PVI (Pilot Vehicle Interface) 131, 149
posture 98, 123-124, 150
prefrontal 70, 76, 80
primary/secondary task inversion 11
proficiency check 49, 54-55, 57
PVT (psychomotor vigilance task) 115-116

quality assurance 125-126, 155

rail alerting system 106, 108-109
reaction time 107-108
recognition of stressors 32, 35, 37, 54, 56-57
recurrent training 38
resistance to change 145-146, 153
resource utilisation 17, 32, 131-132, 142-143, 145, 147, 149, 151
respiration 72, 74, 90, 98, 150
response to accidents 26-27, 131
reticular arousal system 71
road alerting system 104-108
role ambiguity 55-57, 83, 85, 111-112, 114, 116, 132, 139, 141, 147, 149

182

Index

seat-kilometre 85
self focused attention 49
sensory modulation 62
simulator 18, 22-24, 74-75, 77-78, 98, 109, 123, 126, 146, 148
situational awareness 10-12, 14-15, 18-30, 32, 35, 39-46, 81, 100, 132, 140
situational awareness and peripheralisation effects 11, 18, 27-28, 44-46, 81, 100
situational awareness and vigilance 30, 81
situational awareness, definition 14-15
situational awareness in accidents 39-43
situational awareness, training 32, 35, 132, 140
skills-based 32-36, 43, 143, 150
skin conductance 96-97, 99, 105-106, 126-130
skin resistance (see skin conductance) 96-97
sleep cycle 19, 22, 72, 74-75, 77-78, 92, 98, 109, 123, 126, 146, 148
sleep deprivation 22
steering wheel 104, 107-108
strain (stress produced) 35, 48-49, 83
stress 2, 19, 21, 28, 32, 35, 37, 47-57, 73-78, 81-85, 97-100, 113-114, 116, 120, 132, 134, 139, 141-142, 144, 146-147, 149, 153
stress and pilots 48-49, 53-57, 100, 113, 116, 139, 141, 144

stress as error modulator 21
stress management and CRM 32, 35, 56, 146-147, 153
stress mechanisms 47-51, 73-78
stress versus arousal 47, 51-53, 73-74
stressors 32, 37, 48-51, 53-57, 81, 83, 120
subsidiary task 107-108
synergism, teams 34, 145
system view of arousal 51, 52, 89

take-off phase 9, 36, 39-41, 93, 96, 128, 150
task-centred 12-13, 15, 132, 139, 156
teams (in cockpits) 18, 19-20, 34-35, 141, 144-146
tracking error 104, 106-107
training 17, 19-20, 25-26, 28, 32-43, 56, 82, 110, 130, 140, 142-143, 145, 148, 150, 152

ultra-long haul 1, 22, 141, 144
useful stress 21, 51-52

vigilance 1-2, 11, 29-30, 55-57, 59-63, 68-73, 78-85, 89, 91-94, 96, 98-99, 101-102, 104-109, 111-116, 119-126, 130-133, 137-139, 149-150
vigilance and complacency 11
vigilance and monitoring 29-30, 55, 137, 139
vigilance and peripheralisation 56-57, 81-85, 137, 139, 149
vigilance, arousal and stress 63, 73, 78-85, 89, 99, 102, 119-

121, 139, 149
vigilance as a stressor 83, 139
vigilance decrement 30, 59-63, 93, 103, 105
vigilance definition 59
vigilance in cockpits 30, 81, 93-94, 101, 138
vigilance level measurement 105-106, 111, 114, 120-126
vigilance mechanisms 59-63, 68-73, 123
vigilance measurement (see EEG, heart rate, skin conductance 99, 101-102, 104-109, 111-116, 120-126
vigilance performance measurement 107-109, 111-113, 115-116
vigilance theories 59-63

visual attention system 70

warning 8, 111, 114
wind shear 9
workload 9-10, 12, 17-19, 21-22, 34-35, 39, 55, 57, 72, 93, 99-101, 119, 128, 151,
workload and errors 17-19, 21-2, 39, 151
workload and flight phase 9, 151
workload, effort, fatigue 55, 99, 119
workload, high 12, 21-22, 39, 151
workload, low 9, 21-22, 93
workload management 35, 57
workload measurement 72, 93, 99-101, 119, 128